—WORKBOOK

Applied Math
FOR WASTEWATER PLANT OPERATORS

JOANNE KIRKPATRICK PRICE
Training Consultant

CRC PRESS

Boca Raton   London   New York   Washington, D.C.

Visit the CRC Press Web site at www.crcpress.com

No claim to original U.S. Government works
International Standard Book Number 87762-810-6
Printed in the United States of America   7 8 9 0
Printed on acid-free paper

Dedication

This book is dedicated to my family:

To my husband Benton C. Price who was patient and
supportive during the two years it took to write these texts, and
who not only had to carry extra responsibilities during this time,
but also, as a sanitary engineer, provided frequent technical
critique and suggestions.

To our children Lisa, Derek, Kimberly, and Corinne,
who so many times had to pitch in while I was busy writing,
and who frequently had to wait for my attention.

To my mother who has always been so encouraging and who
helped in so many ways throughout the writing process.

To my father, who passed away since the writing of the first
edition, but who, I know, would have had just as instrumental
a role in these books.

To the other members of my family, who have had to put up
with this and many other projects, but who maintain a sense
of humor about it.

Thank you for your love in allowing me to do something
that was important to me.

J.K.P.

Contents

Contents—Cont'd

Contents—Cont'd

Contents—Cont'd

Preface to the Second Edition

The first edition of these texts was written at the conclusion of three and a half years of instruction at Orange Coast College, Costa Mesa, California, for two different water and wastewater technology courses. The fundamental philosophy that governed the writing of these texts was that those who have difficulty in math often do not lack the ability for mathematical calculation, they merely have not learned, or have not been taught, the "language of math." The books, therefore, represent an attempt to bridge the gap between the reasoning processes and the language of math that exists for students who have difficulty in mathematics.

In the years since the first edition, I have continued to consider ways in which the texts could be improved. In this regard, I researched several topics including how people learn (learning styles, etc.), how the brain functions in storing and retrieving information, and the fundamentals of memory systems. Many of the changes incorporated in this second edition are a result of this research.

Two features of this second edition are of particular importance:

- the **skills check section** provided at the beginning of every basic math chapter

- a **grouping of similar types of calculations** in the applied math texts

The skills check feature of the basic math text enables the student to pinpoint the areas of math weakness, and thereby customizes the instruction to the needs of the individual student.

The first six chapters of each applied math text include calculations grouped by type of problem. These chapters have been included so that students could see the common thread in a variety of seemingly different calculations.

The changes incorporated in this second edition were field-tested during a three-year period in which I taught a water and wastewater mathematics course for Palomar Community College, San Marcos, California.

Written comments or suggestions regarding the improvement of any section of these texts or workbooks will be greatly appreciated by the author.

Joanne Kirkpatrick Price

Acknowledgments

"From the original planning of a book to its completion, the continued encouragement and support that the author receives is instrumental to the success of the book." This quote from the acknowledgments page of the first edition of these texts is even more true of the second edition.

First Edition

Those who assisted during the development of the first edition are: Walter S. Johnson and Benton C. Price, who reviewed both texts for content and made valuable suggestions for improvements; Silas Bruce, with whom the author team-taught for two and a half years, and who has a down-to-earth way of presenting wastewater concepts; Mariann Pape, Samuel R. Peterson and Robert B. Moore of Orange Coast College, Costa Mesa, California, and Jim Catania and Wayne Rodgers of the California State Water Resources Control Board, all of whom provided much needed support during the writing of the first edition.

The first edition was typed by Margaret Dionis, who completed the typing task with grace and style. Adele B. Reese, my mother, proofed both books from cover to cover and Robert V. Reese, my father, drew all diagrams (by hand) shown in both books.

Second Edition

The second edition was an even greater undertaking due to many additional calculations and because of the complex layout required. I would first like to acknowledge and thank Laurie Pilz, who did the computer work for all three texts and the two workbooks. Her skill, patience, and most of all perseverance has been instrumental in providing this new format for the texts. Her husband, Herb Pilz, helped in the original format design and he assisted frequently regarding questions of graphics design and computer software.

Those who provided technical review of various portions of the texts include Benton C. Price, Kenneth D. Kerri, Lynn Marshall, Wyatt Troxel and Mike Hoover. Their comments and suggestions are appreciated and have improved the current edition.

Many thanks also to the staff of the Fallbrook Sanitary District, Fallbrook, California, especially Virginia Grossman, Nancy Hector, Joyce Shand, Mike Page, and Weldon Platt for the numerous times questions were directed their way during the writing of these texts.

The staff of Technomic Publishing Company, Inc., also provided much advice and support during the writing of these texts. First, Melvyn Kohudic, President of Technomic Publishing Company, contacted me several times over the last few years, suggesting that the texts be revised. It was his gentle nudging that finally got the revision underway. Joseph Eckenrode helped work out some of the details in the initial stages and was a constant source of encouragement. Jeff Perini was copy editor for the texts. His keen attention to detail has been of great benefit to the final product. Leo Motter had the arduous task of final proof reading.

I wish to thank all my friends, but especially those in our Bible study group (Gene and Judy Rau, Floyd and Juanita Miller, Dick and Althea Birchall, and Mark and Penny Gray) and our neighbors, Herb and Laurie Pilz, who have all had to live with this project as it progressed slowly chapter by chapter, but who remained a source of strength and support when the project sometimes seemed overwhelming.

Lastly, the many students who have been in my classes or seminars over the years have had no small part in the final form these books have taken. The format and content of these texts is in response to their questions, problems, and successes over the years.

To all of these I extend my heartfelt thanks.

How To Use These Books

The *Mathematics for Water and Wastewater Treatment Plant Operators* series includes three texts and two workbooks:

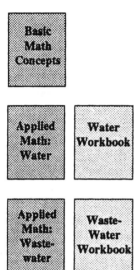

- Basic Math Concepts for Water and Wastewater Plant Operators

- Applied Math for Water Plant Operators

- Workbook—Applied Math for Water Plant Operators

- Applied Math for Wastewater Plant Operators

- Workbook—Applied Math for Wastewater Plant Operators

Basic Math Concepts

All the basic math you will need to become adept in water and wastewater calculations has been included in the Basic Math Concepts text. This section has been expanded considerably from the basic math included in the first edition. For this reason, students are provided with more methods by which they may solve the problems.

Many people have weak areas in their math skills. It is therefore advisable to take the skills test at the beginning of each chapter in the basic math book to pinpoint areas that require review or study. If possible, it is best to resolve these weak areas before beginning either of the applied math texts. However, when this is not possible, the Basic Math Concepts text can be used as a reference resource for the applied math texts. For example, when making a calculation that includes tank volume, you may wish to refer to the basic math section on volumes.

Applied Math Texts and Workbooks

The applied math texts and workbooks are companion volumes. There is one set for water treatment plant operators and another for wastewater treatment plant operators. Each applied math text has two sections:

- Chapters 1 through 6 present various calculations **grouped by type of math problem**. Perhaps 70 percent of all water and wastewater calculations are represented by these six types. Chapter 7 groups various types of pumping problems into a single chapter. The calculations presented in these seven chapters are common to the water and wastewater fields and have therefore been included in both applied math texts.

  Since the calculations described in Chapters 1 through 6 represent the heart of water and wastewater treatment math, if possible, it is advisable that you master these general types of calculations before continuing with other calculations. Once completed, a review of these calculations in subsequent chapters will further strengthen your math skills.

- The remaining chapters in each applied math text include calculations **grouped by unit processes**. The calculations are presented in the order of the flow through a plant. Some of the calculations included in these chapters are not incorporated in Chapters 1 through 7, since they do not fall into any general problem-type grouping. These chapters are particularly suited for use in a classroom or seminar setting, where the math instruction must parallel unit process instruction.

The workbooks support the applied math texts section by section. They have also been vastly expanded in this edition so that the student can build strength in each type of calculation. A detailed answer key has been provided for all problems. The workbook pages have been perforated so that they may be used in a classroom setting as hand-in assignments. The pages have also been hole-punched so that the student may retain the pages in a notebook when they are returned.

The workbooks may be useful in preparing for a certification exam. However, because theses texts include both fundamental and advanced calculations, and because the requirements for each certification level vary somewhat from state to state, it is advisable that you <u>first determine the types of problems to be covered in your exam</u>, then focus on those types of calculations in these texts.

1 *Applied Volume Calculations*

PRACTICE PROBLEMS 1.1: Tank Volume Calculations

1. The diameter of a tank is 80 ft. If the water depth is 30 ft, what is the volume of water in the tank, in gallons?

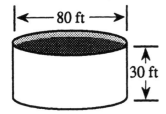

ANS_____

2. The dimensions of a tank are given below. Calculate the cubic feet volume of the tank.

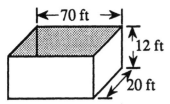

ANS_____

3. A tank 25 ft wide and 80 ft long is filled with water to a depth of 13 ft. What is the volume of water in the tank (in gal)?

ANS_____

4. What is the volume of water in a tank, in gallons, if the tank is 15 ft wide, 30 ft long, and contains water to a depth of 10 ft?

ANS_____

5. Given the tank diameter and depth shown below, calculate the volume of water in the tank, in gallons.

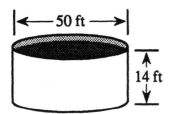

ANS_____

PRACTICE PROBLEMS 1.2: Channel or Pipeline Capacity Calculations

1. What is the cubic feet volume of water in the section of rectangular channel shown below?

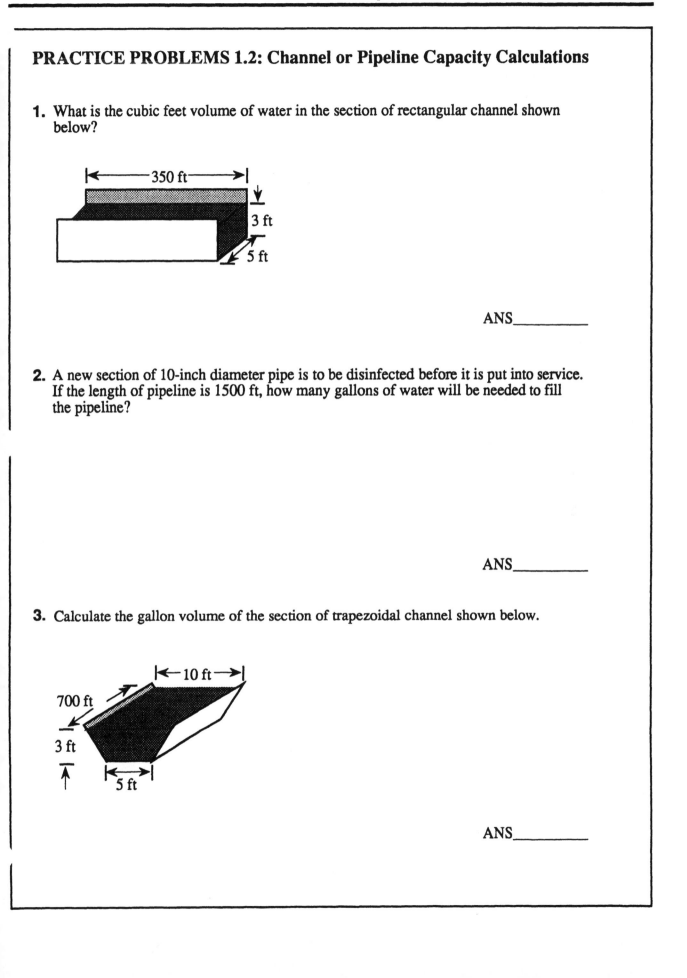

ANS_____

2. A new section of 10-inch diameter pipe is to be disinfected before it is put into service. If the length of pipeline is 1500 ft, how many gallons of water will be needed to fill the pipeline?

ANS_____

3. Calculate the gallon volume of the section of trapezoidal channel shown below.

ANS_____

4. A section of 6-inch diameter pipeline is to be filled with chlorinated water for disinfection. If 1778 ft of pipeline is to be disinfected, how many gallons of chlorinated water will be required?

ANS_____

5. What is the volume of water (in gal) for the 1000-ft section of channel shown below?

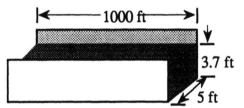

ANS_____

PRACTICE PROBLEMS 1.3: Other Volume Calculations

1. A trench is to be excavated that is 3 ft wide, 3.5 ft deep, and 600 ft long. What is the cubic yards volume of the trench?

ANS_____

2. A pond is 6 ft deep. Given the dimensions as shown below, calculate the cu ft volume of the pond.

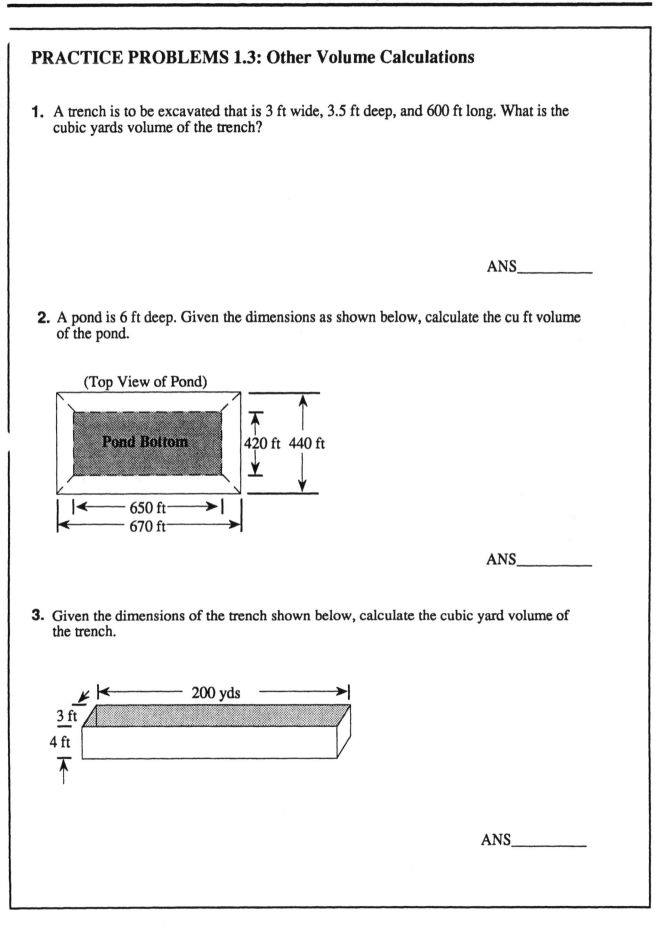

(Top View of Pond)

Pond Bottom

420 ft   440 ft

650 ft

670 ft

ANS_____

3. Given the dimensions of the trench shown below, calculate the cubic yard volume of the trench.

200 yds

3 ft

4 ft

ANS_____

4. Calculate the cu ft volume of the oxidation ditch shown below. The cross section of the ditch is trapezoidal.

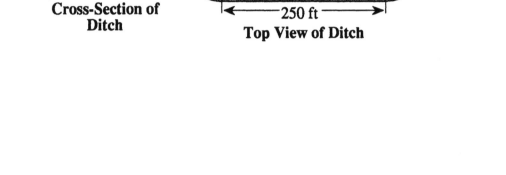

Cross-Section of Ditch

Top View of Ditch

ANS_____

5. A trench is 250 yards long, 2 ft wide and 2 ft deep. What is the cubic feet volume of the trench?

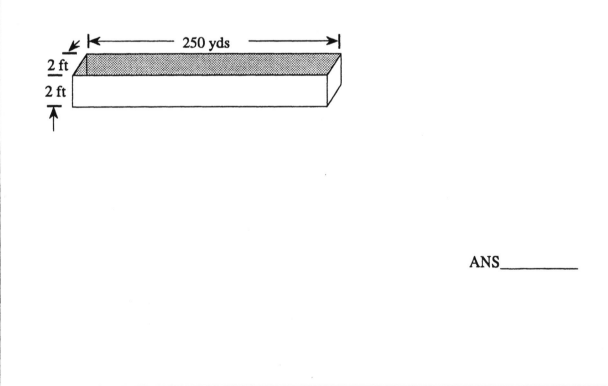

ANS_____

Chapter 1—Achievement Test

1. What is the cubic feet volume of water in the rectangular channel shown below?

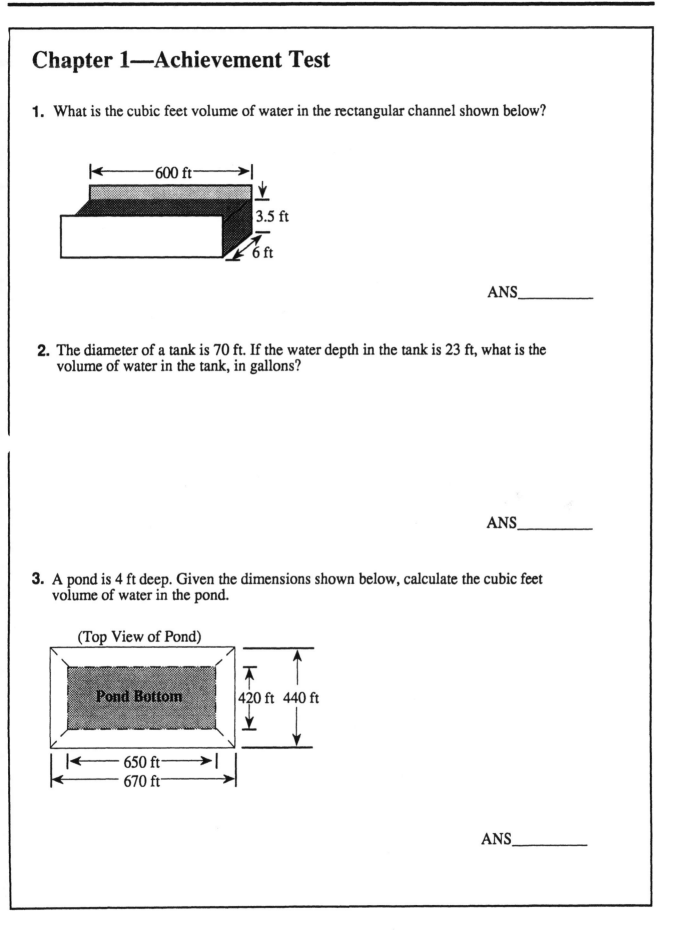

ANS_____

2. The diameter of a tank is 70 ft. If the water depth in the tank is 23 ft, what is the volume of water in the tank, in gallons?

ANS_____

3. A pond is 4 ft deep. Given the dimensions shown below, calculate the cubic feet volume of water in the pond.

4. The dimensions of a tank are given below. Calculate the cubic feet volume of the tank.

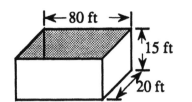

ANS_____

5. A new section of 8-inch diameter pipe is to be filled with water for testing. If the length of pipeline is 3500 ft, how many gallons of water will be needed to fill the pipeline?

ANS_____

6. A trench is 300 yards long, 2 ft wide, and 2.5 ft deep. What is the cubic feet volume of the trench?

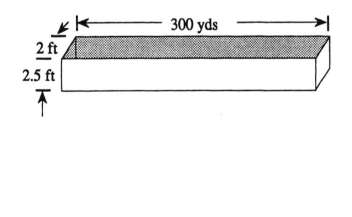

ANS_____

Chapter 1—Achievement Test (Cont'd)

7. A trench is to be excavated. If the trench is 2.5 ft wide, 3 ft deep and 1500 ft long, what is the cubic yards volume of the trench?

ANS_____

8. Calculate the maximum gallon capacity of the section of trapezoidal channel shown below.

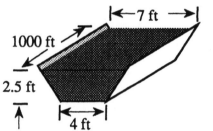

ANS_____

9. A tank is 20 ft wide and 60 ft long. If the tank contains water to a depth of 13 ft, how many gallons of water are in the tank?

ANS_____

10. What is the volume of water (in gal) contained in a 2000-ft section of channel if the channel is 6 ft wide and the water depth is 3.7 ft?

ANS_____

11. Calculate the cu ft capacity of the oxidation ditch shown below. The cross-section of the ditch is trapezoidal. (Round the circumference length to the nearest foot.)

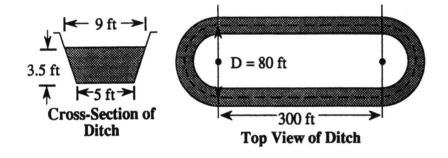

Cross-Section of Ditch

Top View of Ditch

ANS_____

12. Given the diameter and water depth shown below, calculate the volume of water in the tank, in gallons.

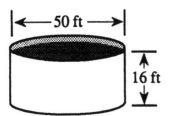

ANS_____

2 *Flow and Velocity Calculations*

PRACTICE PROBLEMS 2.1: Instantaneous Flow Rates

1. A channel 42 inches wide has water flowing to a depth of 2.6 ft. If the velocity of the water is 2.2 fps, what is the cfm flow in channel?

ANS_____

2. A tank is 15 ft long and 10 ft wide. With the discharge valve closed, the influent to the tank causes the water level to rise 0.7 feet in one minute. What is the gpm flow to the tank?

ANS_____

3. A trapezoidal channel is 3.5 ft wide at the bottom and 5.5 ft wide at the water surface. The water depth is 38 inches. If the flow velocity through the channel is 125 ft/min, what is the cfm flow rate through the channel?

ANS_____

4. A 6-inch diameter pipeline has water flowing at a velocity of 2.6 fps. What is the gpm flow rate through the pipeline? Assume the pipe is flowing full. (Round to the nearest tenth.)

ANS_____

5. A pump discharges into a 2-ft diameter barrel. If the water level in the barrel rises 26 inches in 30 seconds, what is the gpm flow into the barrel?

ANS_____

6. A 10-inch diameter pipeline has water flowing at a velocity of 3.2 fps. What is the gpm flow rate through the pipeline if the water is flowing at a depth of 5 inches?

ANS_____

PRACTICE PROBLEMS 2.2: Velocity Calculations

1. A channel has a rectangular cross section. The channel is 5 ft wide with water flowing to a depth of 2.3 ft. If the flow rate through the channel is 13,400 gpm, what is the velocity of the water in the channel (ft/sec)? (Round to the nearest tenth.)

ANS_____

2. An 8-inch diameter pipe flowing full delivers 537 gpm. What is the velocity of flow in the pipeline (ft/sec)? (Round to the nearest tenth.)

ANS_____

3. A fluorescent dye is used to estimate the velocity of flow in a sewer. The dye is injected into the water at one manhole and the travel time to the next manhole 500 ft away is noted. The dye first appears at the downstream manhole in 195 seconds. The dye continues to be visible until the total elapsed time is 221 seconds. What is the ft/sec velocity of flow through the pipeline? (Round to the nearest tenth.)

ANS_____

4. The velocity in a 10-inch diameter pipeline is 2.6 ft/sec. If the 10-inch pipeline flows into an 8-inch diameter pipeline, what is the velocity in the 8-inch pipeline in ft/sec.?

ANS_____

5. A float travels 400 ft in a channel in 1 min 28 sec. What is the estimated velocity in the channel (ft/sec)? (Round to the nearest tenth.)

ANS_____

6. The velocity in a 8-inch diameter pipe is 3.6 ft/sec. If the flow then travels through a 10-inch diameter section of pipeline, what is the ft/sec velocity in the 10-inch pipeline? (Round to the nearest tenth.)

ANS_____

PRACTICE PROBLEMS 2.3: Average Flow Rates

1. The following flows were recorded for the week: Monday—4.6 MGD;
Tuesday—5.2 MGD; Wednesday—5.3 MGD; Thursday—4.9 MGD;
Friday—5.4 MGD; Saturday—5.1 MGD; Sunday—4.8 MGD. What was the
average daily flow rate for the week?

ANS_____

2. The totalizer reading for the month of November was 117.3 MG. What was the
average daily flow (ADF) for the month of November? (Round to the nearest tenth.)

ANS_____

3. The following flows were recorded for the months of April, May, and June:
April—125.6 MG; May—142.4 MG; June—160.2 MG. What was the average daily
flow for this three-month period? (Round to the nearest tenth.)

ANS_____

4. The total flow for one day at a plant was 3,140,000 gallons. What was the average gpm flow for that day?

ANS_____

PRACTICE PROBLEMS 2.4: Flow Conversions

1. Express a flow of 5 cfs in terms of gpm.

ANS_____

2. What is 38 gps expressed as gpd?

ANS_____

3. Convert a flow of 4,270,000 gpd to cfm.

ANS_____

4. What is 5.6 MGD expressed as cfs? (Round to the nearest tenth.)

ANS_____

5. Express 423,690 cfd as gpm.

ANS_____

6. Convert 2730 gpm to gpd.

ANS_____

Chapter 2—Achievement Test

1. A channel has a rectangular cross section. The channel is 6 ft wide with water flowing to a depth of 2.6 ft. If the flow rate through the channel is 15,500 gpm, what is the velocity of the water in the channel (ft/sec)? (Round to the nearest tenth.)

ANS_____

2. The following flows were recorded for the week: Monday—4.1 MGD; Tuesday—3.4 MGD; Wednesday—3.9 MGD; Thursday—4.6 MGD; Friday—3.2 MGD; Saturday—4.9 MGD; Sunday—3.7 MGD. What was the average daily flow rate for the week?

ANS_____

3. A channel 50 inches wide has water flowing to a depth of 3.2 ft. If the velocity of the water is 3.9 fps, what is the cfm flow in the channel?

ANS_____

4. The following flows were recorded for the months of June, July, and August: June—105.2 MG; July—129.6 MG; August—142.8 MG. What was the average daily flow for this three-month period? (Round to the nearest tenth.)

ANS_____

5. A tank is 10 ft by 10 ft. With the discharge valve closed, the influent to the tank causes the water level to rise 8 inches in one minute. What is the gpm flow to the tank?

ANS_____

6. An 8-inch diameter pipe flowing full delivers 490 gpm. What is the ft/sec velocity of flow in the pipeline?

ANS_____

7. Express a flow of 8 cfs in terms of gpm.

ANS_____

8. The totalizer reading for the month of October was 127.6 MG. What was the average daily flow (ADF) for the month of October? (Round to the nearest tenth.)

ANS_____

Chapter 2—Achievement Test Cont'd

9. What is 4.8 MGD expressed as cfs? (Round to the nearest tenth.)

ANS_____

10. A pump discharges into a 2-ft diameter barrel. If the water level in the barrel rises 18 inches in 30 seconds, what is the gpm flow into the barrel?

ANS_____

11. Convert a flow of 1,780,000 gpd to cfm.

ANS_____

12. A 6-inch diameter pipeline has water flowing at a velocity of 2.7 fps. What is the gpm flow rate through the pipeline? (Assume the pipe is flowing full.)

ANS_____

13. A fluorescent dye is used to estimate the velocity of flow in a sewer. The dye is injected into the water at one manhole and the travel time to the next manhole 300 ft away is noted. The dye first appears at the downstream manhole in 77 seconds. The dye continues to be visible until there is a total elapsed time of 95 seconds. What is the ft/sec velocity of flow through the pipeline?

ANS_____

14. The velocity in a 10-inch pipeline is 2.4 ft/sec. If the 10-inch pipeline flows into an 8-inch diameter pipeline, what is the ft/sec velocity in the 8-inch pipeline? (Round to the nearest tenth.)

ANS_____

15. Convert 2150 gpm to gpd.

ANS_____

16. The total flow for one day at a plant was 4,620,000 gallons. What was the average gpm flow for that day?

ANS_____

3 *Milligrams Per Liter to Pounds Per Day Calculations*

PRACTICE PROBLEMS 3.1: Chemical Dosage Calculations

1. Determine the chlorinator setting (lbs/day) needed to treat a flow of 5.1 MGD with a chlorine dose of 2.3 mg/L.

ANS_____

2. To dechlorinate a wastewater, sulfur dioxide is to be applied at a level 3 mg/L more than the chlorine residual. What should the sulfonator feed rate be (lbs/day) for a flow of 3.8 MGD with a chlorine residual of 2.9 mg/L?

ANS_____

3. A total chlorine dosage of 8 mg/L is required to treat a particular water. If the flow is 1.6 MGD and the hypochlorite has 65% available chlorine, how many lbs/day of hypochlorite will be required?

ANS_____

4. What should the chlorinator setting be (lbs/day) to treat a flow of 4.6 MGD if the chlorine demand is 8.5 mg/L and a chlorine residual of 2 mg/L is desired?

ANS_____

5. The chlorine dosage at a plant is 4.1 mg/L. If the flow rate is 6,140,000 gpd, what is the chlorine feed rate in lbs/day?

ANS_____

6. A storage tank is to be disinfected with 50 mg/L of chlorine. If the tank holds 85,000 gallons, how many pounds of chlorine (gas) will be needed?

ANS_____

7. To neutralize a sour digester, one pound of lime is to be added for every pound of volatile acids in the digester liquor. If the digester contains 224,000 gal of sludge with a volatile acid (VA) level of 2,120 mg/L, how many pounds of lime should be added?

ANS_____

8. A flow of 0.72 MGD requires a chlorine dosage of 8 mg/L. If the hypochlorite has 65% available chlorine, how many lbs/day of hypochlorite will be required?

ANS_____

PRACTICE PROBLEMS 3.2: Loading Calculations—BOD, COD, and SS

1. The suspended solids concentration of the wastewater entering the primary system is 425 mg/L. If the plant flow is 1,620,000 gpd, how many lbs/day suspended solids enter the primary system?

ANS_____

2. Calculate the BOD loading (lbs/day) on a stream if the secondary effluent flow is 2.98 MGD and the BOD of the secondary effluent is 26 mg/L.

ANS_____

3. The daily flow to a trickling filter is 5,340,000 gpd. If the BOD content of the trickling filter influent is 280 mg/L, how many lbs/day BOD enter the trickling filter?

ANS_____

4. The flow to an aeration tank is 2460 gpm. If the COD concentration of the water is 135 mg/L, how many pounds of COD are applied to the aeration tank daily? (Round the MGD flow to the nearest hundredth.)

ANS_____

5. The daily flow to a trickling filter is 2290 gpm with a BOD concentration of 295 mg/L. How many lbs of BOD are applied to the trickling filter daily? (Round the MGD flow to the nearest hundredth.)

ANS_____

PRACTICE PROBLEMS 3.3: BOD and SS Removal, lbs/day

1. If 148 mg/L suspended solids are removed by a primary clarifier, how many lbs/day suspended solids are removed when the flow is 5.2 MGD?

ANS_____

2. The flow to a primary clarifier is 1.89 MGD. If the influent to the clarifier has a suspended solids concentration of 315 mg/L and the primary effluent has 126 mg/L SS, how many lbs/day suspended solids are removed by the clarifier?

ANS_____

3. The flow to a trickling filter is 4,790,000 gpd. If the primary effluent has a BOD concentration of 160 mg/L and the trickling filter effluent has a BOD concentration of 24 mg/L, how many pounds of BOD are removed daily?

ANS_____

4. A primary clarifier receives a flow of 2.37 MGD with a suspended solids concentration of 387 mg/L. If the clarifier effluent has a suspended solids concentration of 166 mg/L, how many pounds of suspended solids are removed daily?

ANS_____

5. The flow to the trickling filter is 4,140,000 gpd with a BOD concentration of 215 mg/L. If the trickling filter effluent has a BOD concentration of 97 mg/L, how many lbs/day BOD are removed by the trickling filter?

ANS_____

PRACTICE PROBLEMS 3.4: Pounds of Solids Under Aeration

1. An aeration tank has a volume of 350,000 gallons. If the mixed liquor suspended solids concentration is 2140 mg/L, how many pounds of suspended solids are in the aeration tank?

ANS_____

2. The aeration tank of a conventional activated sludge plant has a mixed liquor volatile suspended solids concentration of 1960 mg/L. If the aeration tank is 110 ft long, 35 ft wide, and has wastewater to a depth of 13 ft, how many pounds of MLVSS are under aeration? (Round tank volume to the nearest ten thousand.)

ANS_____

3. The volume of an oxidation ditch is 24,500 cubic feet. If the MLVSS concentration is 2960 mg/L, how many pounds of volatile solids are under aeration? (Round ditch volume to the nearest ten thousand.)

ANS_____

4. An aeration tank is 120 ft long and 45 ft wide. The operating depth is 15 ft. If the mixed liquor suspended solids concentration is 2440 mg/L, how many pounds of mixed liquor suspended solids are under aeration? (Round tank volume to the nearest ten thousand.)

ANS_____

5. An aeration tank is 110 ft long and 40 ft wide. The depth of wastewater in the tank is 15 ft. If the tank contains an MLSS concentration of 2890 mg/L, how many lbs of MLSS are under aeration? (Round tank volume to the nearest ten thousand.)

ANS_____

PRACTICE PROBLEMS 3.5: WAS Pumping Rate Calculations

1. The WAS suspended solids concentration is 6210 mg/L. If 5300 lbs/day solids are to be wasted, what must the WAS pumping rate be, in MGD? (Round MGD flow to the nearest hundredth.)

ANS_____

2. The WAS suspended solids concentration is 5970 mg/L. If 4600 lbs/day solids are to be wasted, (a) What must the WAS pumping rate be, in MGD? (Round MGD flow to the nearest hundredth.) (b) What is this rate expressed in gpm?

ANS_____

3. It has been determined that 6090 lbs/day of solids must be removed from the secondary system. If the RAS SS concentration is 6540 mg/L, what must be the WAS pumping rate, in gpm? (Round MGD flow to the nearest hundredth.)

ANS_____

4. The RAS suspended solids concentration is 6280 mg/*L*. If a total of 7400 lbs/day solids are to be wasted, what should the WAS pumping rate be, in gpm? (Round MGD flow to the nearest hundredth.)

ANS_____

5. A total of 5700 lbs/day of solids must be removed from the secondary system. If the RAS SS concentration is 7140 mg/*L*, what must be the WAS pumping rate, in gpm? (Round MGD flow to the nearest hundredth.)

ANS_____

Chapter 3—Achievement Test

1. Determine the chlorinator setting (lbs/day) required to treat a flow of 3,820,000 gpd with a chlorine dose of 2.4 mg/L.

ANS_____

2. Calculate the BOD loading (lbs/day) on a stream if the secondary effluent flow is 2.05 MGD and the BOD of the secondary effluent is 15 mg/L.

ANS_____

3. The flow to a primary clarifier is 4.6 MGD. If the influent to the clarifier has a suspended solids concentration of 307 mg/L and the primary effluent suspended solids concentration is 122 mg/L, how many lbs/day suspended solids are removed by the clarifier?

ANS_____

4. What should the chlorinator setting be (lbs/day) to treat a flow of 5.6 MGD if the chlorine demand is 7.9 mg/L and a chlorine residual of 2 mg/L is desired?

ANS_____

5. The suspended solids concentration of the wastewater entering the primary system is 315 mg/*L*. If the plant flow is 3.7 MGD, how many lbs/day suspended solids enter the primary system?

ANS_____

6. A total chlorine dosage of 12 mg/*L* is required to treat a particular water. If the flow is 2.8 MGD and the hypochlorite has 65% available chlorine, how many lbs/day of hypochlorite will be required?

ANS_____

7. A primary clarifier receives a flow of 3.22 MGD with a suspended solids concentration of 340 mg/*L*. If the clarifier effluent has a suspended solids concentration of 150 mg/*L*, how many pounds of suspended solids are removed daily?

ANS_____

8. A storage tank is to be disinfected with 50 mg/*L* of chlorine. If the tank holds 80,000 gallons, how many pounds of chlorine gas will be needed?

ANS_____

Chapter 3—Achievement Test Cont'd

9. An aeration tank is 100 ft long and 40 ft wide. The operating depth is 13 ft. If the mixed liquor suspended solids concentration is 2610 mg/L, how many pounds of mixed liquor suspended solids are under aeration? (Round the tank volume to the nearest ten thousand.)

ANS_____

10. The WAS suspended solids concentration is 5980 mg/L. If 5540 lbs/day solids are to be wasted, what must the WAS pumping rate be, in MGD? (Round to the nearest hundredth MGD.)

ANS_____

11. The flow to an aeration tank is 2200 gpm. If the COD concentration of the water is 125 mg/L, how many pounds COD enter the aeration tank daily?

ANS_____

12. The daily flow to a trickling filter is 2190 gpm. If the BOD concentration of the trickling filter influent is 235 mg/L, how many lbs/day BOD are applied to the trickling filter?

ANS_____

13. The 1.9 MGD influent to the secondary system has a BOD concentration of 230 mg/L. The secondary effluent contains 25 mg/L BOD. How many pounds of BOD are removed each day by the secondary system?

ANS_____

14. The chlorine feed rate at a plant is 340 lbs/day. If the flow is 5,100,000 gpd, what is this dosage expressed in mg/L?

ANS_____

15. It has been determined that 6240 lbs/day solids must be removed from the secondary system. If the RAS SS concentration is 5800 mg/L, what must be the WAS pumping rate, in gpm? (Round the MGD flow to the nearest hundredth.)

ANS_____

4 *Loading Rate Calculations*

PRACTICE PROBLEMS 4.1: Hydraulic Loading Rate Calculations

1. A trickling filter 90 ft in diameter treats a primary effluent flow of 2.3 MGD. If the recirculated flow to the clarifier is 0.7 MGD, what is the hydraulic loading on the trickling filter in gpd/sq ft?

ANS_____

2. The flow to an 80-ft diameter trickling filter is 2,670,000 gpd. The recirculated flow is 1,650,000 gpd. At this flow rate, what is the hydraulic loading rate in gpd/sq ft?

ANS_____

3. A rotating biological contactor treats a flow of 3.6 MGD. The manufacturer data indicates a media surface area of 850,000 sq ft. What is the hydraulic loading rate on the RBC in gpd/sq ft? (Round answer to the nearest tenth.)

ANS_____

4. A pond receives a flow of 1,980,000 gpd. If the surface area of the pond is 15 acres, what is the hydraulic loading in in./day? (Round answer to the nearest tenth.)

ANS_____

5. What is the hydraulic loading rate in gpd/sq ft to an 85-ft diameter trickling filter if the primary effluent flow to the trickling filter is 3,780,000 gpd, and the recirculated flow is 1,300,000 gpd?

ANS_____

6. A 20-acre pond receives a flow of 4.3 acre-feet/day. What is the hydraulic loading on the pond in in./day?

ANS_____

PRACTICE PROBLEMS 4.2: Surface Overflow Rate Calculations

1. A sedimentation tank 80 ft by 30 ft receives a flow of 2.14 MGD. What is the surface overflow rate in gpd/sq ft?

ANS_____

2. A circular clarifier has a diameter of 70 ft. If the primary effluent flow is 2.62 MGD, what is the surface overflow rate in gpd/sq ft?

ANS_____

3. A sedimentation tank is 100 ft long and 40 ft wide. If the flow to the tank is 3.28 MGD what is the surface overflow rate in gpd/sq ft?

ANS_____

4. The primary effluent flow to a clarifier is 1.48 MGD. If the sedimentation tank is 20 ft wide and 80 ft long, what is the surface overflow rate of the clarifier in gpd/sq ft?

ANS_____

5. The flow to a circular clarifier is 2.36 MGD. If the diameter of the clarifier is 60 ft, what is the surface overflow rate in gpd/sq ft?

ANS_____

PRACTICE PROBLEMS 4.3: Filtration Rate Calculations

1. A filter 30 ft by 25 ft receives a flow of 2150 gpm. What is the filtration rate in gpm/sq ft? (Round to the nearest tenth.)

ANS_____

2. A filter 40 ft by 20 ft receives a flow rate of 3080 gpm. What is the filtration rate in gpm/sq ft?

ANS_____

3. A filter 25 ft by 50 ft receives a flow of 2000 gpm. What is the filtration rate in gpm/sq ft?

ANS_____

4. A filter 45 ft by 25 ft treats a flow of 2.2 MGD. What is the filtration rate in gpm/sq ft? (Round to the nearest tenth.)

ANS_____

5. A filter has a surface area of 875 sq ft. If the flow treated is 2975 gpm, what is the filtration rate in gpm/sq ft?

ANS_____

PRACTICE PROBLEMS 4.4: Backwash Rate Calculations

1. A filter 15 ft by 15 ft has a backwash flow rate of 4950 gpm. What is the filter backwash rate in gpm/sq ft?

ANS_____

2. A filter 25 ft by 15 ft has a backwash flow rate of 5100 gpm.. What is the filter backwash rate in gpm/sq ft?

ANS_____

3. A filter is 20 ft by 15 ft. If the backwash flow rate is 3300 gpm, what is the filter backwash rate in gpm/sq ft?

ANS_____

4. A filter 20 ft by 30 ft backwashes at a rate of 3200 gpm. What is this backwash rate expressed as gpm/sq ft?

ANS_____

5. The backwash flow rate for a filter is 3800 gpm. If the filter is 20 ft by 20 ft, what is the backwash rate expressed as gpm/sq ft?

ANS_____

PRACTICE PROBLEMS 4.5: Unit Filter Run Volume Calculations

1. The total water filtered during a filter run is 3,890,000 gallons. If the filter is 15 ft by 40 ft, what is the unit filter run volume (UFRV) in gal/sq ft?

ANS_____

2. The total water filtered during a filter run (between backwashes) is 1,680,000 gallons. If the filter is 15 ft by 15 ft, what is the UFRV in gal/sq ft?

ANS_____

3. A filter 20 ft by 25 ft filters a total of 3.96 MG during the filter run. What is the unit filter run volume in gal/sq ft?

ANS_____

4. The total water filtered between backwashes is 1,339,200 gal. If the length of the filter is 15 ft and the width is 12 ft, what is the unit filter run volume in gal/sq ft?

ANS_____

5. A filter is 30 ft by 25 ft. If the total water filtered between backwashes is 5,625,000 gallons, what is the UFRV in gal/sq ft?

ANS_____

PRACTICE PROBLEMS 4.6: Weir Overflow Rate Calculations

1. A rectangular clarifier has a total of 157 ft of weir. What is the weir overflow rate in gpd/ft when the flow is 1,397,000 gpd?

ANS_____

2. A circular clarifier receives a flow of 2.32 MGD. If the diameter of the weir is 60 ft, what is the weir overflow rate in gpd/ft ?

ANS_____

3. A rectangular clarifier has a total of 235 ft of weir. What is the weir overflow rate in gpd/ft when the flow is 2.6 MGD?

ANS_____

4. The flow rate to a clarifier is 1200 gpm. If the diameter of the weir is 70 ft, what is the weir overflow rate in gpd/ft ?

ANS_____

5. A rectangular sedimentation basin has a total weir length of 188 ft. If the flow to the basin is 4.06 MGD, what is the weir loading rate in gpm/ft?

ANS_____

PRACTICE PROBLEMS 4.7: Organic Loading Rate Calculations

1. A trickling filter 80 ft in diameter with a media depth of 5 feet receives a flow of 2,240,000 gpd. If the BOD concentration of the primary effluent is 215 mg/L, what is the organic loading on the trickling filter in lbs BOD/day/1000 cu ft? (Round media volume to nearest hundred.)

ANS_____

2. The flow to a 3.7-acre wastewater pond is 122,000 gpd. The influent BOD concentration is 175 mg/L. What is the organic loading to the pond in lbs BOD/day/ac?

ANS_____

3. An 85-ft diameter trickling filter with a media depth of 6 ft receives a primary effluent flow of 2,960,000 gpd with a BOD of 125 mg/L. What is the organic loading on the trickling filter in lbs BOD/day/1000 cu ft? (Round media volume to the nearest hundred.)

ANS_____

4. A rotating biological contactor (RBC) receives a flow of 2.15 MGD. If the soluble BOD of the influent wastewater to the RBC is 130 mg/*L* and the surface area of the media is 800,000 sq ft, what is the organic loading rate in lbs BOD/day/1000 sq ft?

ANS_____

5. A 90-ft diameter trickling filter with a media depth of 4 ft receives a primary effluent flow of 3.4 MGD. If the BOD concentration of the wastewater flow to the trickling filter is 140 mg/*L*, what is the organic loading rate in lbs BOD/day/1000 cu ft?(Round media volume to nearest hundred.)

ANS_____

PRACTICE PROBLEMS 4.8: Food/Microorganism Ratio Calculations

1. An activated sludge aeration tank receives a primary effluent flow of 3,350,000 gpd with a BOD of 207 mg/L. The mixed liquor volatile suspended solids is 1950 mg/L and the aeration tank volume is 400,000 gallons. What is the current F/M ratio?(Round to the nearest tenth.)

ANS_____

2. The volume of an aeration tank is 270,000 gallons. The mixed liquor suspended solids is 1690 mg/L. If the aeration tank receives a primary effluent flow of 3,150,000 gpd with a BOD of 192 mg/L, what is the F/M ratio? (Round to the nearest tenth.)

ANS_____

3. The desired F/M ratio at a particular activated sludge plant is 0.7 lbs COD/1 lb mixed liquor volatile suspended solids. If the 2.26 MGD primary effluent flow has a COD of 147 mg/L how many lbs of MLVSS should be maintained?

ANS_____

4. An activated sludge plant receives a flow of 1,920,000 gpd with a COD concentration of 155 mg/L. The aeration tank volume is 245,000 gallons and the MLVSS is 1880 mg/L. What is the current F/M ratio?(Round to the nearest tenth.)

ANS_____

5. The flow to an aeration tank is 2,940,000 gpd, with a BOD content of 185 mg/L. If the aeration tank is 100 ft long, 40 ft wide, has wastewater to a depth of 15 ft, and the desired F/M ratio is 0.5, what is the desired MLVSS concentration (mg/L) in the aeration tank? (Round tank volume to the nearest ten thousand.)

ANS_____

PRACTICE PROBLEMS 4.9: Solids Loading Rate Calculations

1. A secondary clarifier is 80 ft in diameter and receives a combined primary effluent (P.E.) and return activated sludge (RAS) flow of 3.75 MGD. If the MLSS concentration in the aerator is 2780 mg/L, what is the solids loading rate on the secondary clarifier in lbs/day/sq ft? (Round to the nearest tenth.)

ANS_____

2. A secondary clarifier, 75 ft in diameter, receives a primary effluent flow of 2.95 MGD and a return sludge flow of 1.12 MGD. If the MLSS concentration is 2950 mg/L, what is the solids loading rate on the clarifier in lbs/day/sq ft? (Round to the nearest tenth.)

ANS_____

3. The desired solids loading rate for a 60-ft diameter clarifier is 28 lbs/day/sq ft. If the total flow to the clarifier is 3,590,000 gpd (P.E. + RAS flows), what is the desired MLSS concentration?

ANS_____

4. A secondary clarifier 50 ft in diameter receives a primary effluent flow of 2,450,000 gpd and a return sludge flow of 750,000 gpd. If the MLSS concentration is 2180 mg/L, what is the solids loading rate on the clarifier in lbs/day/sq ft? (Round to the nearest tenth.)

ANS_____

5. The desired solids loading rate for a 55-ft diameter clarifier is 20 lbs/day/sq ft. If the total flow to the clarifier is 2,960,000 gpd (P.E. + RAS flows), what is the desired MLSS concentration?

ANS_____

PRACTICE PROBLEMS 4.10: Digester Loading Rate Calculations

1. A digester receives a total of 11,650 lbs/day volatile solids. If the digester volume is 32,600 cu ft, what is the digester loading in lbs VS added/day/cu ft? (Round to the nearest hundredth.)

ANS_____

2. A digester 50 ft in diameter with a water depth of 20 ft receives 122,000 lbs/day raw sludge. If the sludge contains 6.5% solids with 70% volatile matter, what is the digester loading in lbs VS added/day/cu ft? (Round to the nearest hundredth.)

ANS_____

3. A digester 45 ft in diameter with a liquid level of 19 ft receives 139,000 lbs/day sludge with 6% total solids and 69% volatile solids. What is the digester loading in lbs VS added/day/cu ft? (Round to the nearest hundredth.)

ANS_____

4. A digester 35 ft in diameter with a liquid level of 15 ft receives 19,200 gpd sludge with 5.3% solids and 68% volatile solids. What is the digester loading in lbs VS/day/cu ft? Assume the sludge weighs 8.34 lbs/gal. (Round to the nearest hundredth.)

ANS_____

5. A digester 40 ft in diameter with a liquid level of 19 ft receives 21,000 gpd sludge with 5.2% total solids and 71% volatile solids. What is the digester loading in lbs VS/day/cu ft? Assume the sludge weighs 8.8 lbs/gal. (Round to the nearest hundredth.)

ANS_____

PRACTICE PROBLEMS 4.11: Digester Volatile Solids Loading Ratio Calculations

1. A total of 1930 lbs/day volatile solids are pumped to a digester. The digester sludge contains a total of 31,200 lbs of volatile solids. What is the volatile solids loading on the digester in lbs VS added/day/lb VS in digester? (Round to the nearest hundredth.)

ANS_____

2. A digester contains a total of 172,700 lbs of sludge that has a total solids content of 5.8% and volatile solids of 65%. If 550 lbs/day volatile solids are added to the digester, what is the volatile solids loading on the digester in lbs VS added/day/lb VS in digester? (Round to the nearest hundredth.)

ANS_____

3. A total of 61,200 lbs/day sludge is pumped to a 110,000-gallon digester. The sludge being pumped to the digester has a total solids content of 5.4% and a volatile solids content of 71%. The sludge in the digester has a solids content of 6.5% with a 57% volatile solids content. What is the volatile solids loading on the digester in lbs VS added/day/lb VS in digester? Assume the sludge in the digester weighs 8.34 lbs/gal. (Round to the nearest hundredth.)

ANS_____

4. A total of 108,000 gal of digested sludge is in a digester. The digested sludge contains 5.8% total solids and 56% volatile solids. If the desired VS loading ratio is 0.08 lbs VS added/day/lb VS under digestion, what is the desired lbs VS/day to enter the digester? Assume the sludge in the digester weighs 8.34 lbs/gal. (Round to the nearest hundredth.)

ANS_____

5. A total of 7,700 gpd sludge is pumped to the digester. The sludge has 4.4% solids with a volatile solids content of 72%. If the desired VS loading ratio is 0.06 lbs VS added/day/lb VS under digestion, how many lbs VS should be in the digester for this volatile solids load? Assume the sludge pumped to the digester weighs 8.34 lbs/gal. (Round to the nearest hundredth.)

ANS_____

PRACTICE PROBLEMS 4.12: Population Loading and Population Equivalent

1. A 4.7-acre wastewater pond serves a population of 1530. What is the population loading on the pond in persons per acre?

ANS_____

2. A wastewater pond serves a population of 3825. If the pond is 9 acres, what is the population loading on the pond?

ANS_____

3. A 372,000-gpd wastewater flow has a BOD concentration of 1710 mg/*L*. Using an average of 0.2 lbs/day BOD/person, what is the population equivalent of this wastewater flow?

ANS_____

4. A wastewater pond is designed to serve a population of 5000. If the desired population loading is 350 people per acre, how many acres of pond will be required? (Round the answer to the nearest tenth.)

ANS_____

5. A 98,000-gpd wastewater flow has a BOD content of 2190 mg/*L*. Using an average of 0.2 lbs/day BOD/person, what is the population equivalent of this flow?

ANS_____

Chapter 4—Achievement Test

1. A circular clarifier has a diameter of 75 ft. If the primary effluent flow is 2.14 MGD, what is the surface overflow rate in gpd/sq ft?

ANS_____

2. A filter has a square foot area of 180 sq ft. If the flow rate to the filter is 2940 gpm, what is this filter backwash rate expressed as gpm/sq ft? (Round to the nearest tenth.)

ANS_____

3. The flow rate to a circular clarifier is 1,990,000 gpd. If the diameter of the weir is 75 ft, what is the weir overflow rate in gpd/ft?

ANS_____

4. A trickling filter, 80 ft in diameter, treats a primary effluent flow of 2.6 MGD. If the recirculated flow to the clarifier is 0.5 MGD, what is the hydraulic loading on the trickling filter in gpd/sq ft?

ANS_____

5. The desired F/M ratio at an activated sludge plant is 0.6 lbs BOD/day/lb mixed liquor volatile suspended solids. If the 1.8-MGD primary effluent flow has a BOD of 156 mg/L, how many lbs of MLVSS should be maintained in the aeration tank?

ANS_____

6. A digester contains a total of 175,000 lbs sludge that has a total solids content of 6.2% and volatile solids of 68%. If 400 lbs/day volatile solids are added to the digester, what is the volatile solids loading on the digester in lbs/day VS added/lb VS in the digester? (Round to the nearest hundredth.)

ANS_____

7. A secondary clarifier is 75 ft in diameter and receives a combined primary effluent (P.E.) and return activated sludge (RAS) flow of 3.45 MGD. If the MLSS concentration in the aerator is 2640 mg/L, what is the solids loading rate on the secondary clarifier in lbs/day/sq ft?

ANS_____

8. A digester, 60 ft in diameter with a water depth of 22 ft, receives 110,000 lbs/day raw sludge. If the sludge contains 6.8% solids with 70% volatile solids, what is the digester loading in lbs VS added/day/cu ft volume? (Round to the nearest hundredth.)

ANS_____

Chapter 4—Achievement Test Cont'd

9. A 20-acre pond receives a flow of 3.95 acre-feet/day. What is the hydraulic loading on the pond in in./day? (Round to the nearest tenth.)

ANS_____

10. The flow to an aeration tank is 3,154,000 gpd with a BOD content of 172 mg/L. If the aeration tank is 90 ft long, 40 ft wide, has wastewater to a depth of 14 ft, and the desired F/M ratio is 0.5, what is the desired MLVSS concentration (mg/L) in the aeration tank? (Round the tank volume to the nearest thousand.)

ANS_____

11. What is the solids loading rate for a 70-ft diameter clarifier in lbs/day/sq ft if the total flow to the clarifier is 4,100,000 (P.E. + RAS flows) and the MLSS concentration is 3180 mg/L?

ANS_____

12. A sedimentation tank 90 ft by 30 ft receives a flow of 2.1 MGD. What is the surface overflow rate in gpd/sq ft?

ANS_____

13. The total water filtered during a filter run (between backwashes) is 1,740,000 gallons, If the filter is 20 ft by 15 ft, what is the unit filter run volume (UFRV) in gallons/sq ft?

ANS_____

14. The volume of an aeration tank is 290,000 gallons. The mixed liquor volatile suspended solids is 1890 mg/*L*. If the aeration tank receives a primary effluent flow of 2,680,000 gpd with a COD of 140 mg/*L*, what is the F/M ratio? (Round to the nearest tenth.)

ANS_____

15. A total of 23,800 gallons of digested sludge is in a digester. The digested sludge contains 5.7% solids and 58% volatile solids. To maintain a desired VS loading ratio of 0.07 lbs VS added/day/lb VS under digestion, what is the desired lbs VS /day loading to the digester? (Assume sludge in the digester weighs 8.34 lbs/gal.)

ANS_____

16. The flow to a filter is 4.32 MGD. If the filter is 40 ft by 25 ft, what is the filter loading rate in gpm/sq ft?

ANS_____

Chapter 20—Achievement Test Cont'd

17. An 80-ft diameter trickling filter with a media depth of 4 ft receives a primary effluent flow of 3.1 MGD with a BOD concentration of 110 mg/L. What is the organic loading on the filter in lbs BOD/day/1000 cu ft?

ANS_____

18. A circular clarifier receives a flow of 2.46 MGD. If the diameter of the weir is 70 ft, what is the weir overflow rate in gpd/ft?

ANS_____

19. A 5.2-acre wastewater pond serves a population of 1800. What is the population loading on the pond (people/acre)?

ANS_____

20. A rotating biological contactor (RBC) receives a flow of 2.41 MGD. If the soluble BOD of the influent wastewater to the RBC is 120 mg/L and the surface area of the media is 750,000 sq ft, what is the organic loading rate in lbs Sol. BOD/day/1000 sq ft?

ANS_____

21. A filter 40 ft by 20 ft treats a flow of 2.72 MGD. What is the filter loading rate in gpm/sq ft?

ANS_____

5 *Detention and Retention Times Calculations*

PRACTICE PROBLEMS 5.1: Detention Time Calculations

1. A flocculation basin is 7 ft deep, 15 ft wide, and 30 ft long. If the flow through the basin is 1.35 MGD, what is the detention time in minutes?

ANS_____

2. The flow to a sedimentation tank 75 ft long, 25 ft wide, and 10 ft deep is 1.6 MGD. What is the detention time in the tank in hours?

ANS_____

3. A basin, 4 ft by 5 ft, is to be filled to the 2.5-ft level. If the flow to the tank is 5 gpm, how long will it take to fill the tank (in hours)? (Round to the nearest tenth.)

ANS_____

4. The flow rate to a circular clarifier is 4.87 MGD. If the clarifier is 70 ft in diameter with water to a depth of 12 ft, what is the detention time, in hours? (Round to the nearest tenth.)

ANS_____

5. A waste treatment pond is operated at a depth of 5 feet. The average width of the pond is 400 ft and the average length is 550 ft. If the flow to the pond is 219,400 gpd, what is the detention time, in days?

ANS_____

PRACTICE PROBLEMS 5.2: Sludge Age Calculations

1. An aeration tank has a total of 11,900 lbs of mixed liquor suspended solids. If a total of 2627 lbs/day suspended solids enter the aerator in the primary effluent flow, what is the sludge age in the aeration tank? (Round to the nearest tenth.)

ANS_____

2. An aeration tank is 100 ft long, 25 ft wide with wastewater to a depth of 15 ft. The mixed liquor suspended solids concentration is 2670 mg/L. If the primary effluent flow is 958,000 gpd with a suspended solids concentration of 128 mg/L, what is the sludge age in the aeration tank? (Round to the nearest tenth.)

ANS_____

3. An aeration tank contains 220,000 gallons of wastewater. The MLSS is 2960 mg/L. If the primary effluent flow is 1.46 MGD with a suspended solids concentration of 84 mg/L, what is the sludge age?

ANS_____

4. The 1.25 MGD primary effluent flow to an aeration tank has a suspended solids concentration of 70 mg/L. The aeration tank volume is 195,000 gallons. If a sludge age of 6.5 days is desired, what is the desired MLSS concentration?

ANS_____

5. A sludge age of 5.5 days is desired. Assume 1560 lbs/day suspended solids enter the aeration tank in the primary effluent. To maintain the desired sludge age, how many lbs of MLSS must be maintained in the aeration tank?

ANS_____

PRACTICE PROBLEMS 5.3: Solids Retention Time Calculations

1. An aeration tank has a volume of 270,000 gallons. The final clarifier has a volume of 190,000 gallons. The MLSS concentration in the aeration tank is 3100 mg/L. If a total of 1540 lbs/day SS are wasted and 330 lbs/day SS are in the secondary effluent, what is the solids retention time for the activated sludge system? Use the SRT equation that uses combined aeration tank and final clarifier volumes to estimate system solids. (Round to the nearest tenth.)

ANS_____

2. Determine the solids retention time (SRT) given the data below. Use the SRT equation that uses combined aeration tank and final clarifier volumes to estimate system solids. (Round answer to the nearest tenth.)

Aer. Vol. 225,000 gal
Fin. Clar. Vol. 100,000 gal
P.E. Flow 2.15 MGD
WAS Pumping Rate 18,500 gpd

MLSS 2810 mg/L
WAS SS 5340 mg/L
S.E. SS 15 mg/L

ANS_____

3. Calculate the solids retention time given the data below. Use the SRT equation that uses combined aeration tank and final clarifier volumes to estimate system solids. (Round to the nearest tenth.)

Aer. Tank Vol. 1.2 MG
Fin. Clar. Vol. 0.3 MG
P.E. Flow 2.6 MGD
WAS Pumping Rate 75,000 gpd

MLSS 2440 mg/L
WAS SS 6120 mg/L
S.E. SS 18 mg/L

ANS_____

4. The volume of an aeration tank is 600,000 gal and the final clarifier is 155,000 gal. The desired SRT for the plant is 8 days. The primary effluent flow is 2.4 MGD and the WAS pumping rate is 30,000 gpd. If the WAS SS concentration is 6320 mg/L and the secondary effluent SS concentration is 22 mg/L, what is the desired MLSS concentration in mg/L? (Use the SRT equation that uses combined aeration tank and final clarifier volumes to estimate system solids.)

ANS_____

Chapter 5—Achievement Test

1. The flow to a sedimentation tank 70 ft long, 25 ft wide and 12 ft deep is 1,580,000 gpd. What is the detention time in the tank in hours? (Round to the nearest tenth.)

ANS_____

2. An aeration tank has a total of 12,400 lbs of mixed liquor suspended solids. If a total of 2750 lbs/day suspended solids enter the aeration tank in the primary effluent flow, what is the sludge age in the aeration tank? (Round to the nearest tenth.)

ANS_____

3. An aeration tank has a volume of 290,000 gallons. The final clarifier has a volume of 180,000 gallons. The MLSS concentration in the aeration tank is 2950 mg/L. If a total of 1620 lbs/day suspended solids are wasted and 310 lbs/day suspended solids are in the secondary effluent, what is the solids retention time for the activated sludge system? Use the combined aeration tank and clarifier volume to calculate system solids. (Round to the nearest tenth.)

ANS_____

4. The flow through a flocculation basin is 1.62 MGD. If the basin is 35 ft long, 15 ft wide, and 8 ft deep, what is the detention time in minutes?

ANS_____

5. Determine the solids retention time (SRT) given the data below. Use the combined aeration tank and clarifier volume to calculate system solids. (Round to the nearest tenth.)

Aer. Vol.—200,000 gal
Fin. Clar. Vol.—125,000 gal
P.E. Flow—2,300,000 gpd
WAS Pumping Rate—19,200 gpd

MLSS—2740 mg/L
WAS SS—5910 mg/L
SE SS—16 mg/L

ANS_____

6. The mixed liquor suspended solids concentration in an aeration tank is 3140 mg/L. The aeration tank contains 320,000 gallons. If the primary effluent flow is 2,240,000 gpd with a suspended solids concentration of 105 mg/L, what is the sludge age? (Round to the nearest tenth.)

ANS_____

Chapter 5—Achievement Test—Cont'd

7. Calculate the solids retention time given the following data:
   Use the SRT equation that includes the combined aeration tank and secondary clarifier volumes to estimate system solids. (Round to the nearest tenth.)

   Aer. Tank Vol.—1.3 MG   MLSS—2370 mg/L
   Fin. Clar. Vol.—0.4 MG   WAS SS—6210 mg/L
   P.E. Flow—2.72 MGD   SE SS—20 mg/L
   WAS Pumping Rate—69,400 gpd

   ANS_____

8. An aeration tank is 90 ft long, 30 ft wide, with wastewater to a depth of 12 ft. The mixed liquor suspended solids concentration is 2580 mg/L. If the influent flow to the aeration tank is 810,000 gpd with a suspended solids concentration of 145 mg/L, what is the sludge age in the aeration tank? (Round tank volume to the nearest ten thousand and round the sludge age answer to the nearest tenth.)

   ANS_____

9. A tank 5 ft in diameter is to be filled to the 3-ft level. If the flow to the tank is 10 gpm, how long will it take to fill the tank (in min)?

   ANS_____

10. A sludge age of 5 days is desired. The suspended solids concentration of the 1.38 MGD influent flow to the aeration tank is 135 mg/L. To maintain the desired sludge age, how many pounds of MLSS must be maintained in the aeration tank?

ANS_____

11. The average width of a pond is 300 ft and the average length is 480 ft. The depth is 5 ft. If the flow to the pond is 195,000 gpd, what is the detention time in days? (Round to the nearest tenth.)

ANS_____

12. The volume of an aeration tank is 540,000 gal and the volume of the final clarifier is 170,000 gal. The desired SRT for the plant is 8 days. The primary effluent flow is 2,830,000 gpd and the WAS pumping rate is 36,000 gpd. If the WAS SS concentration is 6240 mg/L, and the secondary effluent SS concentration is 15 mg/L, what is the desired MLSS concentration in mg/L? (Use the SRT equation that includes the combined aeration tank and secondary clarifier volumes to estimate system solids.)

ANS_____

6 *Efficiency and Other Percent Calculations*

PRACTICE PROBLEMS 6.1: Unit Process Efficiency Calculations

1. The suspended solids concentration entering a trickling filter is 120 mg/L. If the suspended solids concentration in the trickling filter effluent is 22 mg/L, what is suspended solids removal efficiency of the trickling filter?

ANS_____

2. The BOD concentration of the raw wastewater at an activated sludge plant is 245 mg/L. If the BOD concentration of the final effluent is 15 mg/L, what is the overall efficiency of the plant in BOD removal?

ANS_____

3. The influent flow to a waste treatment pond has a BOD content of 270 mg/L. If the pond effluent has a BOD content of 62 mg/L, what is the BOD removal efficiency of the pond?

ANS_____

4. The influent of a primary clarifier has a BOD content of 230 mg/*L*. If 95 mg/*L* BOD are removed, what is the BOD removal efficiency?

ANS_____

5. The suspended solids concentration of the primary clarifier influent is 305 mg/*L*. If the suspended solids concentration of the primary effluent is 137 mg/*L*, what is the suspended solids removal efficiency?

ANS_____

PRACTICE PROBLEMS 6.2: Percent Solids and Sludge Pumping Rate Calculations

1. A total of 3610 gallons of sludge are pumped to a digester. If the sludge has a 5.8% solids content, how many lbs/day solids are pumped to the digester? (Assume the sludge weighs 8.34 lbs/gal.)

ANS_____

2. The total weight of a sludge sample is 13.05 grams (sludge sample only, not the dish). If the weight of the solids after drying is 0.68 grams, what is the percent total solids of the sludge?

ANS_____

3. A total of 1430 lbs/day SS are removed from a primary clarifier and pumped to a sludge thickener. If the sludge has a solids content of 3.2%, how many lbs/day sludge is this?

ANS_____

4. It is anticipated that 265 lbs/day SS will be pumped from the primary clarifier of a new plant. If the primary clarifier sludge has a solids content of 4.4%, how many gpd sludge will be pumped from the clarifier? (Assume a sludge weight of 8.34 lbs/gal.)

ANS_____

5. A total of 286,000 lbs/day sludge is pumped from a primary clarifier to a sludge thickener. If the total solids content of the sludge is 3.4%, how many lbs/day total solids are sent to the thickener?

ANS_____

PRACTICE PROBLEMS 6.3: Mixing Different Percent Solids Sludges Calculations

1. A primary sludge flow of 3000 gpd with a solids content of 4.5% is mixed with a thickened secondary sludge flow of 4200 gpd that has a solids content of 3.8%. What is the percent solids content of the mixed sludge flow? Assume the density of both sludges is 8.34 lbs/gal. (Round to the nearest tenth percent.)

ANS_____

2. Primary and thickened secondary sludges are to be mixed and sent to the digester. The 8200-gpd primary sludge has a solids content of 5.2% and the 7000-gpd thickened secondary sludge has a solids content of 4.2%. What would be the percent solids content of the mixed sludge? (Assume the density of both sludges is 8.34 lbs/gal.)

ANS_____

3. A 4840-gpd primary sludge has a solids content of 4.9%. The 5200-gpd thickened secondary sludge has a solids content of 3.6%. If the sludges were blended, what would be the percent solids content of the mixed sludge? (Assume the density of both sludges is 8.34 lbs/gal.)

ANS_____

4. A primary sludge flow of 9010 gpd with a solids content of 4.2% is mixed with a thickened secondary sludge flow of 10,760 gpd with a 6.5% solids content. What is the percent solids of the combined sludge flow? (Assume the density of both sludges is 8.34 lbs/gal.)

ANS_____

PRACTICE PROBLEMS 6.4: Percent Volatile Solids Calculations

1. If 3340 lbs/day solids with a volatile solids content of 70% are sent to the digester, how many lbs/day volatile solids are sent to the digester?

ANS_____

2. A total of 4070 gpd of sludge is to be pumped to the digester. If the sludge has a 7% solids content with 72% volatile solids, how many lbs/day volatile solids are pumped to the digester? (Assume the sludge weighs 8.34 lbs/gal.)

ANS_____

3. How many lbs/day volatile solids are pumped to the digester if a total of 6400 gpd of sludge is to be pumped to the digester? The sludge has a 6.5% solids content of which 67% are volatile solids. (Assume the sludge weighs 8.34 lbs/gal.)

ANS_____

4. A 6.8% sludge has a volatile solids content of 66%. If 24,510 lbs/day of sludge are pumped to the digester, how many lbs/day of volatile solids are pumped to the digester?

ANS_____

5. A sludge has a solids content of 6.2% and a volatile solids content of 71%. If 2530 gpd of sludge are pumped to the digester, how many lbs/day volatile solids are pumped to the digester? (Assume the sludge weighs 8.34 lbs/gal.)

ANS_____

PRACTICE PROBLEMS 6.5: Seed Sludge Based on Percent Digester Volume Calculations

1. A digester has a capacity of 280,000 gallons. If the digester seed sludge is to be 18% of the digester capacity, how many gallons of seed sludge will be required?

ANS_____

2. A digester 70 ft in diameter has a side wall water depth of 21 ft. If the digester seed sludge is to be 20% of the digester capacity, how many gallons of seed sludge will be required?

ANS_____

3. A 50-ft diameter digester has a typical water depth of 21 ft. If the seed sludge to be used is 15% of the tank capacity, how many gallons of seed sludge will be required?

ANS_____

4.   A 60-ft diameter digester has a typical side wall water depth of 23 ft. If 87,500 gallons of seed sludge are to be used in starting up the digester, what percent of the digester volume will be seed sludge?

ANS_____

PRACTICE PROBLEMS 6.6: Solution Strength Calculations

1. A total of 2 lbs of chemical is dissolved in 80 lbs of solution. What is the percent strength, by weight, of the solution?

ANS_____

2. If 6 ounces of dry polymer are added to 8 gallons of water, what is the percent strength (by weight) of the polymer solution? (Round pounds dry polymer and pounds water to the nearest tenth and round the answer to the nearest tenth.)

ANS_____

3. How many pounds of dry polymer must be added to 50 gallons of water to make a 1.5% (by weight) polymer solution? (Round the answer to the nearest tenth.)

ANS_____

4. If 500 grams of dry polymer are dissolved in 8 gallons of water, what percent strength is the solution? (1 gram = 0.0022 lbs)

ANS_____

5. How many grams of chemical must be dissolved in 5 gallons of water to make a 1.8% solution? (1 lb = 454 grams)

ANS_____

PRACTICE PROBLEMS 6.7: Mixing Different Percent Strength Solutions Calculations

1. If 15 lbs of a 9% strength solution are mixed with 100 lbs of a 1% strength solution, what is the percent strength of the solution mixture? (Round to the nearest tenth.)

ANS_____

2. If 10 lbs of a 12% strength solution are mixed with 330 lbs of a 0.3% strength solution, what is the percent strength of the solution mixture? (Round to the nearest tenth.)

ANS_____

3. If 20 lbs of an 12% strength solution is mixed with 445 lbs of a 0.5% strength solution, what is the percent strength of the solution mixture? (Round to the nearest tenth.)

ANS_____

4. If 10 gallons of a 11% strength solution are added to 60 gallons of a 0.1% strength solution, what is the percent strength of the solution mixture? Assume the 11% strength solution weighs 9.8 lbs/gal and the 0.1% strength solution weighs 8.34 lbs/gal. (Round to the nearest tenth.)

ANS_____

PRACTICE PROBLEMS 6.8: Pump and Motor Efficiency Calculations

1. The brake horsepower of a pump is 22 hp. If the water horsepower is 17 hp, what is the efficiency of the pump?

ANS_____

2. If the motor horsepower is 50 hp and the brake horsepower is 43 hp, what is the efficiency of the motor?

ANS_____

3. The motor horsepower is 25 hp. If the motor is 89% efficient, what is the brake horsepower?

ANS_____

4. A total of 50 hp is supplied to a motor. If the wire-to-water efficiency of the pump and motor is 62%, what will the whp be?

ANS_____

5. The brake horsepower is 34.4 hp. If the motor is 86% efficient, what is the motor horsepower?

ANS_____

Chapter 6—Achievement Test

1. The BOD concentration of the raw wastewater at an activated sludge plant is 230 mg/L. If the BOD concentration of the final effluent is 21 mg/L, what is the overall efficiency of the plant in BOD removal?

ANS_____

2. A total of 7200 gpd sludge is to be pumped to the digester. If the sludge has a 7% solids content with 68% volatile solids, how many lbs/day volatile solids are pumped to the digester? (Assume the sludge weighs 8.34 lbs/gal.)

ANS_____

3. A primary sludge flow of 2600 gpd with a solids content of 4.2% is mixed with a secondary sludge flow of 3380 gpd with a solids content of 3.5%. What is the percent solids content of the mixed sludge flow? (Assume the weight of both sludges is 8.34 lbs/gal.)

ANS_____

4. A total of 3 pounds of chemical has been dissolved in 92 pounds of solution. What is the percent strength, by weight, of the solution?

ANS_____

5. A digester 80 ft in diameter has a side water depth of 23 ft. If the digester seed sludge is to be 20% of the digester capacity, how many gallons of seed sludge will be required?

ANS_____

6. If 12 lbs of a 10% strength solution are mixed with 400 lbs of a 0.15% strength solution, what is the percent strength of the solution mixture? (Round to the nearest tenth.)

ANS_____

7. The brake horsepower of a pump is 25.5 hp. If the water horsepower is 20.4 hp, what is the efficiency of the pump?

ANS_____

8. The influent flow to a waste treatment pond has a BOD content of 350 mg/L. If the pond effluent has a BOD content of 81 mg/L, what is the BOD removal efficiency of the pond?

ANS_____

Chapter 6—Achievement Test—Cont'd

9. A primary sludge flow of 8700 gpd with a solids content of 4.8% is mixed with a thickened secondary sludge flow of 11,200 gpd with a solids content of 6.2%. What is the percent solids of the combined sludge flow? Assume the weight of both sludges is 8.34 lbs/gal. (Round to the nearest tenth.)

ANS_____

10. How many lbs/day volatile solids are pumped to the digester if a total of 7400 gpd of sludge is pumped to the digester? The sludge has a 6.2% solids content, of which 72% are volatile solids.

ANS_____

11. A total of 35,400 lbs/day of sludge is pumped to the digester. If the sludge has a solids content of 5.7%, how many lbs/day solids are pumped to the digester?

ANS_____

12. How many pounds of dry polymer must be added to 75 gallons of water to make a 0.5% polymer solution?

ANS_____

13. A digester has a capacity of 190,000 gallons. If the digester seed sludge is to be 25% of the digester capacity, how many gallons of seed sludge will be required?

ANS_____

14. If 6800 lbs/day solids with a volatile solids content of 71% are pumped to the digester, how many lbs/day volatile solids are pumped to the digester?

ANS_____

15. If 8 pounds of a 9% strength solution are mixed with 90 pounds of a 1% strength solution, what is the percent strength of the solution mixture? (Round to the nearest tenth.)

ANS_____

16. If the motor horsepower is 50 hp and the brake horsepower is 38 hp, what is the efficiency of the motor?

ANS_____

17. A total of 75 hp is supplied to a motor. If the wire-to-water efficiency of the pump and motor is 67%, what will the water horsepower be?

ANS_____

7 *Pumping Calculations*

PRACTICE PROBLEMS 7.1: Density and Specific Gravity

1. A gallon of solution is weighed. After the weight of the container is subtracted, it is determined that the weight of the solution is 8.9 lbs. What is the density of the solution?

ANS_____

2. The density of a substance is given as 67.1 lbs/cu ft. What is this density expressed as lbs/gal?

ANS_____

3. The density of a liquid is given as 55 lbs/cu ft. What is the specific gravity of the liquid?

ANS_____

4. The specific gravity of a liquid is 1.3. What is the density of that liquid? (The density of water is 8.34 lbs/gal.)

ANS_____

5. You wish to determine the specific gravity of a solution. After weighing a gallon of solution and subtracting the weight of the container, the solution is found to weigh 8.9 lbs. What is the specific gravity of the solution?

ANS_____

PRACTICE PROBLEMS 7.2: Pressure and Force Calculations

1. The object shown below weighs 75 lbs. What is the lbs/sq in. pressure at the surface of contact? (Round to the nearest tenth.)

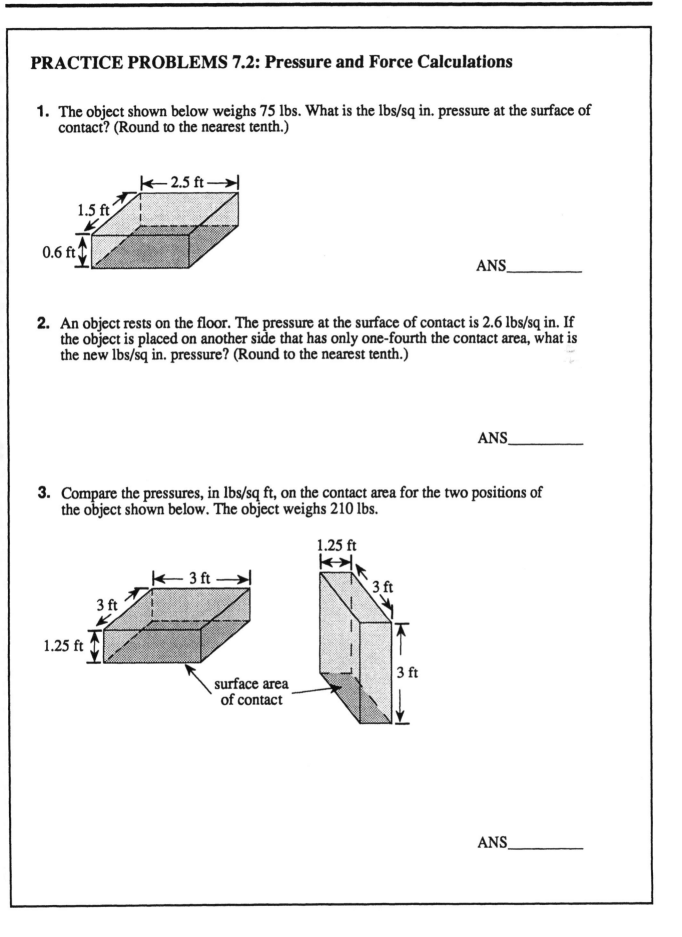

ANS_____

2. An object rests on the floor. The pressure at the surface of contact is 2.6 lbs/sq in. If the object is placed on another side that has only one-fourth the contact area, what is the new lbs/sq in. pressure? (Round to the nearest tenth.)

ANS_____

3. Compare the pressures, in lbs/sq ft, on the contact area for the two positions of the object shown below. The object weighs 210 lbs.

surface area of contact

ANS_____

4. What is the pressure (in lbs/sq ft) at a point 6 ft below the surface of the water? (The density of water is 62.4 lbs/cu ft.)

ANS_____

5. What is the pressure (in psi) at a point 8 ft below the surface?

ANS_____

6. At a point 4.5 ft below the liquid surface, what is the pressure in psi? (The specific gravity of the liquid is 1.4.)

ANS_____

PRACTICE PROBLEMS 7.2: Pressure and Force Calculations—Cont'd

7. What is the total force against the bottom of a tank 20 ft long and 12 ft wide? The water depth is 10 ft. (Round psi to the nearest tenth.)

ANS_____

8. What is the total force exerted on the side of a tank if the tank is 25 ft wide and the water depth is 10 ft?

ANS_____

9. The side of a tank is 15 ft wide. The water depth is 9 ft. At what depth is the center of force against the tank wall?

ANS_____

10. The force applied to the small cylinder of a hydraulic jack is 50 lbs. The diameter of the small cylinder is 8 inches. If the diameter of the large cylinder is 2 ft, what is the total lifting force?

ANS_____

11. A gage reading is 32 psi. What is the absolute pressure at the gage? (Assume sea level atmospheric pressure.)

ANS_____

PRACTICE PROBLEMS 7.3: Head and Head Loss Calculations

1. The elevation of two water surfaces are 322 ft and 239 ft. What is the total static head, in ft?

ANS_____

2. The elevation of two water surfaces are 790 ft and 614 ft. If the friction and minor head losses equal 16 ft, what is the total dynamic head (in ft)?

ANS_____

3. The pump inlet and outlet pressure gage readings are given below. (Pump is off.) If the friction and minor head losses are equal to 12 ft, what is the total dynamic head (in ft)?

95 psi   162 psi

ANS_____

4. A 6-inch diameter pipe has a C-value of 100. When the flow rate is 900 gpm, what is the friction loss for a 1000-ft length of pipe? (Use the table given in Appendix A.)

ANS_____

5. Flow through a 10-inch diameter pipeline is 1450 gpm. The *C*-value is 100. What is the friction loss through a 2000-ft section of pipe? (Use the table given in Appendix A.)

ANS_____

6. Determine the "equivalent length of pipe" for a flow through a gate valve, 1/4 closed, for an 8-inch diameter pipeline. (Use the nomograph given in Appendix B.)

ANS_____

PRACTICE PROBLEMS 7.4: Horsepower Calculations

1. A pump must pump 1500 gpm against a total head of 40 ft. What horsepower is required for this work? (Round to the nearest tenth.)

    ANS_____

2. If 20 hp is supplied to a motor (mhp), what is the bhp and whp if the motor is 85% efficient and the pump is 80% efficient? (Round to the nearest tenth.)

    ANS_____

3. A total of 35 hp is required for a particular pumping application. If the pump efficiency is 85%, what is the brake horsepower required? (Round to the nearest tenth.)

    ANS_____

4. A pump must pump against a total dynamic head of 60 ft at a flow rate of 600 gpm. The liquid to be pumped has a specific gravity of 1.2. What is the water horsepower requirement for this pumping application?

ANS_____

5. The motor horsepower requirement has been calculated to be 45 hp. How many kilowatts electric power does this represent? (1 hp = 746 watts)

ANS_____

6. The motor horsepower requirement has been calculated to be 75 hp. During the week, the pump is in operation a total of 144 hours. Using a power cost of $0.06125/kWh, what would be the power cost that week for the pumping?

ANS_____

PRACTICE PROBLEMS 7.5: Pump Capacity Calculations

1. A wet well is 12 ft long and 10 ft wide. The influent valve to the wet well is closed. If a pump lowers the water level 2.6 ft during a 5-minute pumping test, what is the gpm pumping rate?

ANS_____

2. A pump is discharged into a 55-gallon barrel. If it takes 29 seconds to fill the barrel, what is the pumping rate in gpm?

ANS_____

3. A pump test is conducted for 5 minutes while influent flow continues. During the test, the water level rises 2 inches. If the tank is 8 ft by 10 ft and the influent flow is 400 gpm, what is the pumping rate in gpm?

ANS_____

4. A piston pump discharges a total of 0.75 gal/stroke. If the pump operates at 30 strokes per minute, what is the gpm pumping rate? Assume the piston is 100% efficient and displaces 100% of its volume each stroke. (Round to the nearest tenth.)

ANS_____

5. A sludge pump has a bore of 8 inches and a stroke setting of 4 inches. The pump operates at 40 strokes per minute. If the pump operates a total of 150 minutes during a 24-hour period, what is the gpd pumping rate? Assume the piston is 100% efficient.

ANS_____

Chapter 7—Achievement Test

1. The density of a substance is given as 66 lbs/cu ft. What is this density expressed as lbs/gal?

ANS_____

2. The dimensions of a wet well are 12 ft by 10 ft. The influent valve to the wet well is closed. If a pump lowers the water level 1.5 ft during a 5-minute pumping test, what is the gpm capacity of the pump?

ANS_____

3. The object shown below weighs 90 lbs. What is the lbs/sq in. pressure at the surface of contact?

ANS_____

4. A sludge pump has a bore of 6 inches and a stroke of 2.5 inches. If the pump operates at 55 strokes (or revolutions) per minute, how many gpm are pumped? (Assume the piston is 100% efficient and displaces 100% of its volume each stroke.)

ANS_____

5. A pump test is conducted for 5 minutes. If the water level in the 12 ft long and 10 ft wide wet well drops 1.2 ft during the test, what is the gpm pumping rate? (Assume there is no influent to the wet well during the test.)

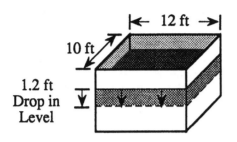

ANS_____

6. A pump must pump 900 gpm against a total head of 70 ft. What horsepower is required for this work? (Round to the nearest tenth.)

ANS_____

7. A sludge pump has a bore of 8 inches and a stroke setting of 3 inches. The pump operates at 50 revolutions per minute. If the pump operates a total of 125 minutes during a 24-hour period, what is the gpd pumping rate? (Assume the piston is 100% efficient.)

ANS_____

8. The specific gravity of a liquid is 1.4. What is the density of that liquid? (The density of water = 8.34 lbs/gal.)

ANS_____

Chapter 7—Achievement Test—Cont'd

9. What is the pressure (in psi) at a point 7 ft below the water surface?

ANS_____

10. The elevations of two water surfaces are 852 ft and 760 ft. If the friction and minor head losses equal 10 ft, what is the total dynamic head (in ft)?

ANS_____

11. What is the head loss through a gate valve (fully open) for a 6-inch diameter pipeline? (Use the nomograph provided in Appendix B.)

ANS_____

12. What is the total force exerted on the side of a tank if the tank is 15 ft wide and the water depth is 9 ft?

ANS_____

13. If 50 hp is supplied to a motor, what is the bhp and whp if the motor is 90% efficient and the pump is 85% efficient?

ANS_____

14. An 8-inch diameter pipe has a *C*-value of 100. When the flow rate is 1000 gpm, what is the friction loss for a 3000-ft length of pipe? (Use the table provided in Appendix A.)

ANS_____

15. The flow through ana 6-inch diameter pipeline is 240 gpm. The *C*-value is 100. What is the friction loss through a 1200-ft length of pipe? (Use the table provided in Appendix A.)

ANS_____

16. The motor horsepower requirement has been calculated to be 60 hp. During the week, the pump is in operation a total of 120 hours. Using a power cost of $0.05626/kWh, what would be the power cost that week for the pumping?

ANS_____

8 *Wastewater Collection and Preliminary Treatment*

PRACTICE PROBLEMS 8.1: Wet Well Capacity

1. What is the gallon capacity of a wet well 10 ft long, 10 ft wide, and 8 ft deep?

ANS_____

2. A wet well is 12 ft long, 10 ft wide, and contains water to a depth of 6 ft. How many gallons of water does it contain?

ANS_____

3. What is the cubic feet capacity of a wet well 8 ft by 8 ft with a maximum depth of 6 ft?

ANS_____

4. The maximum capacity of a wet well is 4787 gallons. If the wet well is 10 ft long and 8 ft wide, what is the maximum depth of water in the wet well?

ANS_____

5. A wet well is 8 ft long and 6 ft wide. If the wet well contains water to a depth of 2.8 ft, what is the volume of water in the wet well, in gallons?

ANS_____

PRACTICE PROBLEMS 8.2: Wet Well Pumping Rate

1. A wet well is 8 ft by 8 ft. During a 5-minute pumping test, with no influent to the well, a pump lowers the water level 1.6 ft. What is the pumping rate in gpm?

ANS_____

2. A wet well is 10 ft by 12 ft. During a 3-minute pumping test, a pump lowers the water level 1.1 ft. What is the gpm pumping rate? (Assume no influent to the well during the pumping test.)

ANS_____

3. The water level in a wet well drops 19 inches during a 3-minute pumping test. There was no influent to the wet well during the pumping test. If the wet well is 8 ft by 6 ft, what is the pumping rate in gpm?

ANS_____

4. During a period when there is no pumping from the wet well, the water level rises 0.7 ft in one minute. If the wet well is 8 ft long and 7 ft wide, what is the gpm flow rate of wastewater entering the wet well?

ANS_____

PRACTICE PROBLEMS 8.3: Screenings Removed

1. A total of 62 gallons of screenings are removed from the wastewater flow during a 24-hour period. What is the screenings removal reported as cu ft/day? (Round to the nearest tenth.)

ANS_____

2. During one week a total of 271 gallons of screenings were removed from the wastewater screens. What was the average screenings removal in cu ft/day? (Round to the nearest tenth.)

ANS_____

3. The flow at a treatment plant is 2.72 MGD. If a total of 4.7 cu ft of screenings are removed during a 24-hour period, what is the screenings removal reported as cu ft/MG? (Round to the nearest tenth.)

ANS_____

4. On a particular day, a treatment plant receives a flow of 4.6 MGD. If 78 gallons of screenings are removed that day, what is the screenings removal expressed as cu ft/MG? (Round to the nearest tenth.)

ANS_____

5. A total of 45 gallons of screenings are removed from the treatment plant during a 24-hour period. If the treatment plant received a flow of 2,170,000 gpd, what is the screenings removal expressed as cu ft/MG? (Round to the nearest tenth.)

ANS_____

PRACTICE PROBLEMS 8.4: Screenings Pit Capacity

1. A screenings pit has a capacity of 500 cu ft. (The pit is actually larger than 500 cu ft to accommodate soil for covering.) If an average of 2.8 cu ft of screenings are removed daily from the wastewater flow, in how many days will the pit be full?

    ANS_____

2. A screenings pit has a capacity of 8 cu yds available for screenings. If the plant removes an average of 1.4 cu ft of screenings per day, in how many days will the pit be filled?

    ANS_____

3. A plant has been averaging a screenings removal of 2.1 cu ft/MG. If the average daily flow is 2.7 MGD, how many days will it take to fill a screenings pit with an available capacity of 290 cu ft?

    ANS_____

4. Suppose you want to use a screenings pit for 120 days. If the screenings removal rate is 3.4 cu ft/day, what is the required screenings pit capacity in cu ft? (Calculate only the capacity for screenings. An additional capacity will be required for cover material.)

ANS_____

PRACTICE PROBLEMS 8.5: Grit Channel Velocity

1. A grit channel is 3 ft wide, with water flowing to a depth of 19 inches. If the flow meter indicates a flow rate of 1750 gpm, what is the velocity of flow through the channel in ft/sec? (Round to the nearest tenth.)

ANS_____

2. A stick in a grit channel travels 25 ft in 31 seconds. What is the estimated velocity in the channel in ft/sec? (Round to the nearest tenth.)

ANS_____

3. The total flow through both channels of a grit channel is 4.1 cfs. If each channel is 2 ft wide and water is flowing to a depth of 13 inches, what is the velocity of flow through the channel in ft/sec? (Round to the nearest tenth.)

ANS_____

4. A stick is placed in a grit channel and flows 35 ft in 31 seconds. What is the estimated velocity in the channel in ft/sec? (Round to the nearest tenth.)

ANS_____

5. The depth of water in a grit channel is 15 inches. The channel is 32 inches wide. If the flow meter indicates a flow of 1120 gpm, what is the velocity of flow through the channel in ft/sec? (Round to the nearest tenth.)

ANS_____

PRACTICE PROBLEMS 8.6: Grit Removal

1. A treatment plant removes 11 cu ft of grit in one day. If the plant flow was 7 MGD, what is this removal expressed as cu ft/MG?

ANS_____

2. The total daily grit removal for a plant is 240 gallons. If the plant flow is 10.3 MGD, how many cubic feet of grit are removed per MG flow? (Round to the nearest tenth.)

ANS_____

3. The average grit removal at a particular treatment plant is 2.8 cu ft/MG. If the monthly average daily flow is 3.6 MGD, how many cu yds of grit would be removed from the wastewater flow during one month? (Assume the month has 30 days.)

ANS_____

4. The monthly average grit removal is 2.1 cu ft/MG. If the average daily flow for the month is 4,120,000 gpd, how many cu yds must be available for grit disposal if the disposal pit is to have a 90-day capacity? (Calculate only the volume required for grit, not cover material.)

ANS_____

PRACTICE PROBLEMS 8.7: Flow Measurement

1. A grit channel 2.5 ft wide has water flowing to a depth of 15 inches. If the velocity through the channel is 0.9 fps, what is the cfs flow rate through the channel? (Round to the nearest tenth.)

ANS_____

2. A grit channel 2 ft wide has water flowing at a velocity of 1.2 fps. If the depth of water is 16 inches, what is the gpd flow rate through the channel? (Assume the flow is steady and continuous.)

ANS_____

3. A grit channel 34 inches wide has water flowing to a depth of 9 inches. If the velocity of the water is 0.85 fps, what is the cfs flow in the channel? (Round to the nearest tenth.)

ANS_____

4. Using the table given below, determine the cfs flow rate through a rectangular weir with end contractions if the feet of head indicated at the staff gage is 0.12 ft and the length of the weir crest is 2 ft.

| Head | LENGTH OF WEIR CREST IN FEET | | | | | |
| ft | 1 | | 1-1/2 | | 2 | |
	cfs	MGD	cfs	MGD	cfs	MGD
0.11	.119	.077	.178	.116	.241	.155
0.12	.135	.087	.204	.132	.273	.177
0.13	.152	.098	.230	.149	.308	.199
0.14	.169	.110	.256	.166	.343	.222

ANS_____

DISCHARGE OF 6-INCH PARSHALL FLUME							
Head ft	cfs	gps	MGD	Head ft	cfs	gps	MGD
0.31	.3238	2.422	.2092	0.41	.5036	3.767	.3255
0.32	.3404	2.547	.2200	0.42	.5231	3.913	.3381
0.33	.3574	2.673	.2310	0.43	.5429	4.062	.3509
0.34	.3746	2.803	.2421	0.44	.5630	4.212	.3639
0.35	.3922	2.934	.2535	0.45	.5834	4.364	.3770
0.36	.4100	3.068	.2650	0.46	.6040	4.518	.3904
0.37	.4282	3.203	.2767	0.47	.6249	4.675	.4038
0.38	.4466	3.341	.2886	0.48	.6460	4.833	.4175
0.39	.4653	3.481	.3007	0.49	.6674	4.993	.4313
0.40	.4843	3.623	.3130	0.50	.6890	5.155	.4453

5. The head measured at the upstream gage on a 6-inch Parshall flume is 0.33 ft. What is the MGD flow through the channel? (Assume there are no submergence conditions in the channel.)

ANS_____

6. What is the cfs flow through a 6-inch Parshall flume if the upstream gage indicates a depth of 0.39 ft? (Assume no submergence condition exists.)

ANS_____

7. The head measured at the upstream gage of a 6-inch Parshall flume is 0.48 ft. What is the gpm flow through the flume? (Assume no submergence condition exists.)

ANS_____

PRACTICE PROBLEMS 8.7: Flow Measurement Calculations Cont'd

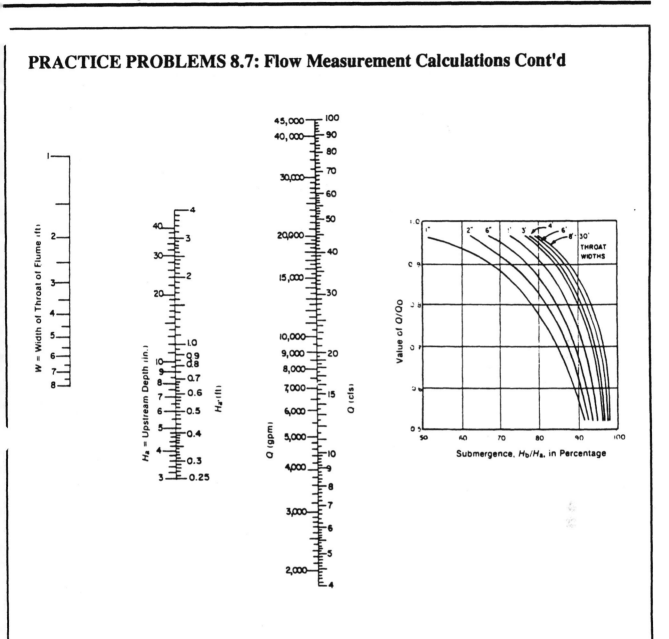

8. What is the cfs flow through a Parshall flume with a throat width of 2.5 ft if the water depth at the upstream gage is 0.9 ft? The downstream depth (H$_b$) is 0.6 ft. (Submergence conditions exist when the percent submergence exceeds 70% for flumes 1-8 ft wide.)

ANS_____

9. What is the cfs flow through a Parshall flume with a throat width of 3 ft if the water depth at the upstream gage is 10 in.? The downstream depth (Hb) is 8.5 in. (Submergence conditions exist when the percent submergence exceeds 70% for flumes 1-8 ft wide.)

ANS_____

Chapter 8—Achievement Test

1. A plant has been averaging a screenings removal of 2.4 cu ft/MG. If the average daily flow is 2,900,000 gpd, how many days will it take to fill a screenings pit that has an available capacity of 260 cu ft?

ANS_____

2. During 7 days a total of 200 gallons of screenings were removed from the wastewater screens. What was the average screenings removal in cu ft/day? (Round to the nearest tenth.)

ANS_____

3. A total of 5.2 cu ft of screenings are removed from the wastewater flow during a 24-hour period. If the flow at the treatment plant is 2,840,000 gpd, what is the screenings removal reported as cu ft/MG? (Round to the nearest tenth.)

ANS_____

4. A screenings pit has a capacity of 10 cubic yards available for screenings. If the plant removes an average of 2.3 cu ft of screenings per day, in how many days will the pit be filled?

ANS_____

5. A float is placed in a channel. If the float travels 35 ft in 28 seconds, what is the estimated velocity in the channel in ft/sec? (Round to the nearest tenth.)

ANS_____

6. Using the table shown below, determine the cfs flow rate through a rectangular weir with end contractions if the feet of head indicated at the staff gage is 0.14 and the length of the weir crest is 1 ft.

Head ft	LENGTH OF WEIR CREST IN FEET					
	1		1-1/2		2	
	cfs	MGD	cfs	MGD	cfs	MGD
0.11	.119	.077	.178	.116	.241	.155
0.12	.135	.087	.204	.132	.273	.177
0.13	.152	.098	.230	.149	.308	.199
0.14	.169	.110	.256	.166	.343	.222

ANS_____

7. What is the MGD flow through a 6-inch Parshall flume if the upstream gage indicates a depth of 0.32 ft? (Assume no submergence condition exists.)

DISCHARGE THROUGH A 6-INCH PARSHALL FLUME							
Head ft	cfs	gps	MGD	Head ft	cfs	gps	MGD
0.31	.3238	2.422	.2092	0.41	.5036	3.767	.3255
0.32	.3404	2.547	.2200	0.42	.5231	3.913	.3381
0.33	.3574	2.673	.2310	0.43	.5429	4.062	.3509
0.34	.3746	2.803	.2421	0.44	.5630	4.212	.3639
0.35	.3922	2.934	.2535	0.45	.5834	4.364	.3770
0.36	.4100	3.068	.2650	0.46	.6040	4.518	.3904
0.37	.4282	3.203	.2767	0.47	.6249	4.675	.4038
0.38	.4466	3.341	.2886	0.48	.6460	4.833	.4175
0.39	.4653	3.481	.3007	0.49	.6674	4.993	.4313
0.40	.4843	3.623	.3130	0.50	.6890	5.155	.4453

ANS_____

Chapter 8—Achievement Test—Cont'd

8. A grit channel 2.5 ft wide has water flowing to a depth of 14 inches. If the velocity of the water is 0.9 fps, what is the cfs flow in the channel? (Round to the nearest tenth.)

    ANS_____

9. The total daily grit removal for a treatment plant is 210 gallons. If the plant flow is 8.6 MGD, how many cubic feet of grit are removed per MG flow? (Round to the nearest tenth.)

    ANS_____

10. A grit channel is 2.5 ft wide with water flowing to a depth of 16 inches. If the flow velocity through the channel is 1.7 ft/sec, what is the gpm flow through the channel?

    ANS_____

11. The average grit removal at a particular treatment plant is 2.2 cu ft/MG. If the monthly average daily flow is 3,540,000 gpd, how many cu yds of grit would be expected to be removed from the wastewater flow during one month? Assume the month has 30 days. (Round to the nearest tenth.)

ANS_____

12. A grit channel 3 ft wide has water flowing to a depth of 11 inches. If the velocity through the channel is 1 fps, what is the cfs flow rate through the channel? (Round to the nearest tenth.)

ANS_____

9 *Sedimentation*

PRACTICE PROBLEMS 9.1: Detention Time

1. A circular clarifier has a capacity of 150,000 gallons. If the flow through the clarifier is 1,720,000 gpd, what is the detention time in hours for the clarifier? (Round to the nearest tenth.)

ANS_____

2. The flow to a sedimentation tank 80 ft long, 20 ft wide, and 12 ft deep is 3.15 MGD. What is the detention time in the tank, in hours? (Round to the nearest tenth.)

ANS_____

3. A circular clarifier receives a flow of 4,340,000 gpd. If the clarifier is 80 ft in diameter and 12 ft deep, what is the clarifier detention time in hours? Assume the flow is steady and continuous. (Round to the nearest tenth.)

ANS_____

4. The flow to a sedimentation tank 70 ft long, 30 ft wide, and 12 ft deep is 3.12 MGD. What is the detention time in hours in the tank? (Round to the nearest tenth.)

ANS_____

PRACTICE PROBLEMS 9.2: Weir Overflow Rate

1. A rectangular clarifier has a total of 110 ft of weir. What is the weir overflow rate in gpd/ft when the flow is 1,463,000 gpd?

ANS_____

2. A circular clarifier receives a flow of 2.95 MGD. If the diameter of the weir is 70 ft, what is the weir overflow rate in gpd/ft?

ANS_____

3. The average flow to a clarifier is 2440 gpm. If the diameter of the weir is 80 ft, what is the weir overflow rate in gpd/ft?

ANS_____

4. The total ft of weir for a clarifier is 188 ft. If the flow to the weir is 1.78 MGD, what is the weir overflow rate in gpd/ft?

ANS_____

PRACTICE PROBLEMS 9.3: Surface Overflow Rate

1. A circular clarifier has a diameter of 60 ft. If the primary clarifier influent flow is 2,890,000 gpd, what is the surface overflow rate in gpd/sq ft?

ANS_____

2. A sedimentation basin 75 ft by 25 ft receives a flow of 2.16 MGD. What is the surface overflow rate in gpd/sq ft?

ANS_____

3. A sedimentation tank is 90 ft long and 40 ft wide. If the flow to the tank is 2,590,000 gpd, what is the surface overflow rate in gpd/sq ft?

ANS_____

4. The average flow to a secondary clarifier is 2580 gpm. What is the surface overflow rate in gpd/sq ft if the secondary clarifier has a diameter of 70 ft?

ANS_____

PRACTICE PROBLEMS 9.4: Solids Loading Rate

1. A secondary clarifier is 75 ft in diameter and receives a combined primary effluent (P.E.) and return activated sludge (RAS) flow of 3.9 MGD. If the MLSS concentration in the aeration tank is 2940 mg/L, what is the solids loading rate on the secondary clarifier in lbs/day/sq ft? (Round to the nearest tenth.)

ANS_____

2. A secondary clarifier, 75 ft in diameter, receives a primary effluent flow of 3.1 MGD and a return sludge flow of 0.9 MGD. If the MLSS concentration is 3140 mg/L, what is the solids loading rate on the clarifier in lbs/day/sq ft? (Round to the nearest tenth.)

ANS_____

3. The MLSS concentration in an aeration tank is 2610 mg/L. The 60-ft diameter secondary clarifier receives a combined primary effluent (P.E.) and return activated sludge (RAS) flow of 3,130,000 gpd. What is the solids loading rate on the secondary clarifier in lbs SS/day/sq ft? (Round to the nearest tenth.)

ANS_____

4. A secondary clarifier, 70 ft in diameter, receives a primary effluent flow of 2,230,000 gpd and a return sludge flow of 640,000 gpd. If the MLSS concentration is 3260 mg/L, what is the solids loading rate on the clarifier in lbs/day/sq ft? (Round to the nearest tenth.)

ANS_____

PRACTICE PROBLEMS 9.5: BOD and SS Removed, lbs/day

1. If 125 mg/L suspended solids are removed by a primary clarifier, how many lbs/day suspended solids are removed when the flow is 5,160,000 gpd?

ANS_____

2. The flow to a primary clarifier is 2,840,000 gpd. If the influent to the clarifier has a suspended solids concentration of 235 mg/L and the primary effluent has 190 mg/L SS, how many lbs/day suspended solids are removed by the clarifier?

ANS_____

3. The flow to a secondary clarifier is 4.17 MGD. If the influent BOD concentration is 195 mg/L and the effluent BOD concentration is 108 mg/L, how many pounds of BOD are removed daily?

ANS_____

4. The flow to a primary clarifier is 960,000 gpd. If the influent to the clarifier has a suspended solids concentration of 310 mg/L and the primary effluent has a suspended solids concentration of 124 mg/L, how many lbs/day suspended solids are removed by the clarifier?

ANS_____

PRACTICE PROBLEMS 9.6: Unit Process Efficiency Calculations

1. The suspended solids entering a primary clarifier is 220 mg/L. If the suspended solids in the primary clarifier effluent is 97 mg/L, what is the suspended solids removal efficiency of the primary clarifier?

ANS_____

2. The suspended solids entering a primary clarifier is 195 mg/L. If the suspended solids in the primary clarifier effluent is 78 mg/L, what is the suspended solids removal efficiency of the primary clarifier?

ANS_____

3. The influent to a primary clarifier has a BOD content of 270 mg/L. If the primary clarifier effluent has a BOD concentration of 59 mg/L, what is the BOD removal efficiency of the primary clarifier?

ANS_____

4. The BOD concentration of a primary clarifier is 310 mg/L. If the primary clarifier effluent BOD concentration is 192 mg/L, what is the BOD removal efficiency of the primary clarifier?

ANS_____

Chapter 9—Achievement Test

1. The flow to a circular clarifier is 3,940,000 gpd. If the clarifier is 75 ft in diameter and 12 ft deep, what is the clarifier detention time in hours? (Round to the nearest tenth.)

ANS_____

2. A circular clarifier has a diameter of 50 ft. If the primary clarifier influent flow is 2,260,000 gpd, what is the surface overflow rate in gpd/sq ft?

ANS_____

3. A rectangular clarifier has a total of 210 ft of weir. What is the weir overflow rate in gpd/ft when the flow is 3,728,000 gpd?

ANS_____

4. A secondary clarifier, 55 ft in diameter, receives a primary effluent flow of 1,887,000 gpd and a return sludge flow of 528,000 gpd. If the MLSS concentration is 2640 mg/L, what is the solids loading rate in lbs/day/sq ft on the clarifier? (Round to the nearest tenth.)

ANS_____

5. A circular primary clarifier has a diameter of 60 ft. If the influent flow to the clarifier is 2.62 MGD, what is the surface overflow rate in gpd/sq ft?

ANS_____

6. A secondary clarifier, 70 ft in diameter, receives a primary effluent flow of 2,740,000 gpd and a return sludge flow of 790,000 gpd. If the mixed liquor suspended solids concentration is 2815 mg/L, what is the solids loading rate in the clarifier in lbs/day/sq ft? (Round to the nearest tenth.)

ANS_____

7. The flow to a secondary clarifier is 5.1 MGD. If the influent BOD concentration is 216 mg/L and the effluent BOD concentration is 103 mg/L, how many lbs/day BOD are removed daily?

ANS_____

8. The flow to a sedimentation tank 80 ft long, 30 ft wide, and 14 ft deep is 4.05 MGD. What is the detention time in the tank, in hours? (Round to the nearest tenth.)

ANS_____

Chapter 9—Achievement Test—Cont'd

9. The average flow to a clarifier is 1920 gpm. If the diameter of the weir is 60 ft, what is the weir overflow rate in gpd/ft?

ANS_____

10. The flow to a secondary clarifier is 4,360,000 gpd. How many pounds of BOD are removed daily if the influent BOD concentration is 182 mg/L and the effluent BOD concentration is 101 mg/L?

ANS_____

11. The flow to a primary clarifier is 3.76 MGD. If the influent to the clarifier has a suspended solids concentration of 230 mg/L and the primary clarifier effluent has a suspended solids concentration of 81 mg/L, how many lbs/day suspended solids are removed by the clarifier?

ANS_____

12. The primary clarifier influent has a BOD concentration of 250 mg/L. If the primary clarifier effluent has a BOD concentration of 68 mg/L, what is the BOD removal efficiency of the primary clarifier?

ANS_____

13. A sedimentation tank is 80 ft long and 35 ft wide. If the flow to the tank is 2,125,000 gpd, what is the surface overflow rate in gpd/sq ft?

ANS_____

10 Trickling Filters

PRACTICE PROBLEMS 10.1: Hydraulic Loading Rate

1. A trickling filter, 75 ft in diameter, treats a primary effluent flow of 640,000 gpd. If the recirculated flow to the trickling filter is 110,000 gpd, what is the hydraulic loading rate in gpd/sq ft on the trickling filter?

ANS_____

2. A high-rate trickling filter receives a flow of 2310 gpm. If the filter has a diameter of 85 ft, what is the hydraulic loading rate in gpd/sq ft on the filter?

ANS_____

3. The total influent flow (including recirculation) to a trickling filter is 1.4 MGD. If the trickling filter is 90 ft in diameter, what is the hydraulic loading rate in gpd/sq ft on the trickling filter?

ANS_____

4. A high rate trickling filter receives a daily flow of 1.9 MGD. What is the hydraulic loading rate in MGD/acre if the filter is 95 ft in diameter? (Round acres to the nearest hundredth. Round the answer to the nearest tenth.)

ANS_____

PRACTICE PROBLEMS 10.2: Organic Loading Rate

1. A trickling filter, 90 ft in diameter with a media depth of 5 ft, receives a flow of 1,200,000 gpd. If the BOD concentration of the primary effluent is 205 mg/L, what is the organic loading on the trickling filter in lbs BOD/day/1000 cu ft? (Round to the nearest tenth.)

ANS_____

2. A 80-ft diameter trickling filter with a media depth of 6 ft receives a primary effluent flow of 3,240,000 gpd with a BOD of 110 mg/L. What is the organic loading on the trickling filter in lbs BOD/day/1000 cu ft? (Round to the nearest tenth.)

ANS_____

3. A trickling filter, 70 ft in diameter, with a media depth of 6 ft receives a flow of 0.8 MGD. If the BOD concentration of the primary effluent is 195 mg/L, what is the organic loading on the trickling filter in lbs BOD/day/1000 cu ft? (Round to the nearest tenth.)

ANS_____

4. A trickling filter has a diameter of 95 feet and a media depth of 6 feet. The primary effluent has a BOD concentration of 130 mg/L. If the total flow to the filter is 1.3 MGD, what is the organic loading in lbs per acre-ft? (Round ac-ft volume to the nearest tenth.)

ANS_____

PRACTICE PROBLEMS 10.3: BOD and SS Removed, lbs/day

1. If 121 mg/*L* suspended solids are removed by a trickling filter, how many lbs/day suspended solids are removed when the flow is 3,178,000 gpd?

ANS_____

2. The flow to a trickling filter is 1.61 MGD. If the primary effluent has a BOD concentration of 240 mg/*L* and the trickling filter effluent has a BOD concentration of 72 mg/*L*, how many pounds of BOD are removed?

ANS_____

3. If 177 mg/*L* of BOD are removed from a trickling filter when the flow to the trickling filter is 2,840,000 gpd, how many lbs/day BOD are removed?

ANS_____

4. The flow to a trickling filter is 5.1 MGD. If the trickling filter effluent has a BOD concentration of 26 mg/*L* and the primary effluent has a BOD concentration of 215 mg/*L*, how many pounds of BOD are removed daily?

ANS_____

PRACTICE PROBLEMS 10.4: Unit Process or Overall Efficiency

1. The suspended solids concentration entering a trickling filter is 146 mg/L. If the suspended solids concentration in the trickling filter effluent is 46 mg/L, what is the suspended solids removal efficiency of the trickling filter?

ANS_____

2. The influent to a primary clarifier has a BOD content of 252 mg/L. The trickling filter effluent BOD is 20 mg/L. What is the BOD removal efficiency of the treatment plant?

ANS_____

3. The suspended solids entering a trickling filter is 197 mg/L. If the suspended solids in the trickling filter effluent is 24 mg/L, what is the suspended solids removal efficiency of the trickling filter?

ANS_____

4. The suspended solids concentration entering a trickling filter is 110 mg/*L*. If 86 mg/*L* suspended solids are removed from the trickling filter, what is the suspended solids removal efficiency of the trickling filter?

ANS_____

PRACTICE PROBLEMS 10.5: Recirculation Ratio

1. A treatment plant receives a flow of 3.2 MGD. If the trickling filter effluent is recirculated at the rate of 3.4 MGD, what is the recirculation ratio? (Round to the nearest tenth.)

ANS_____

2. The influent to the trickling filter is 1.52 MGD. If the recirculated flow is 2.12 MGD, what is the recirculation ratio? (Round to the nearest tenth.)

ANS_____

3. The trickling filter effluent is recirculated at the rate of 3.73 MGD. If the treatment plant receives a flow of 2.62 MGD, what is the recirculation ratio? (Round to the nearest tenth.)

ANS_____

4. A trickling filter has a desired recirculation ratio of 1.4. If the primary effluent flow is 4.4 MGD, what is the desired recirculated flow in MGD? (Round to the nearest tenth.)

ANS_____

Chapter 10—Achievement Test

1. A standard-rate filter, 90 ft in diameter, treats a primary effluent flow of 540,000 gpd. If the recirculated flow to the trickling filter is 120,000 gpd, what is the hydraulic loading rate on the filter in gpd/sq ft?

ANS_____

2. A trickling filter, 85 ft in diameter with a media depth of 5 ft, receives a flow of 1,200,000 gpd. If the BOD concentration of the primary effluent is 160 mg/L, what is the organic loading on the trickling filter in lbs BOD/day/1000 cu ft?

ANS_____

3. If 113 mg/L suspended solids are removed by a trickling filter, how many lbs/day suspended solids are removed when the flow is 2,668,000 gpd?

ANS_____

4. The suspended solids concentration entering a trickling filter is 210 mg/L. If the suspended solids concentration in the trickling filter effluent is 67 mg/L, what is the suspended solids removal efficiency of the trickling filter?

ANS_____

5. The flow to a trickling filter is 1.33 MGD. If the primary effluent has a BOD concentration of 231 mg/L and the trickling filter effluent has a BOD concentration of 83 mg/L, how many pounds of BOD are removed?

ANS_____

6. A high-rate trickling filter receives a combined primary effluent and recirculated flow of 2.75 MGD. If the filter has a diameter of 80 ft, what is the hydraulic loading rate on the filter in gpd/sq ft?

ANS_____

7. The influent of a primary clarifier has a BOD content of 205 mg/L. The trickling filter effluent BOD is 20 mg/L. What is the BOD removal efficiency of the treatment plant?

ANS_____

8. A 80-ft diameter trickling filter with a media depth of 7 ft receives a flow of 2,180,000 gpd. If the BOD concentration of the primary effluent is 139 mg/L, what is the organic loading on the trickling filter in lbs BOD/day/1000 cu ft?

ANS_____

Chapter 10—Achievement Test—Cont'd

9. A trickling filter has a diameter of 90 feet and an average media depth of 7 feet. The primary effluent has a BOD concentration of 140 mg/L. If the total flow to the filter is 1.05 MGD, what is the organic loading in lbs/day per acre-ft? (Round ac-ft volume to the nearest hundredth.)

ANS_____

10. The flow to a trickling filter is 4.11 MGD. If the trickling filter effluent has a BOD concentration of 19 mg/L and the primary effluent has a BOD concentration of 197 mg/L, how many pounds of BOD are removed daily?

ANS_____

11. A treatment plant receives a flow of 3.3 MGD. If the trickling filter effluent is recirculated at the rate of 3.6 MGD, what is the recirculation ratio?

ANS_____

12. A high-rate trickling filter receives a daily flow of 1.7 MGD. What is the hydraulic loading rate in MGD/acre if the filter is 90 ft in diameter?

ANS_____

13. The total influent flow (including recirculation) to a trickling filter is 1.89 MGD. If the trickling filter is 80 ft in diameter, what is the hydraulic loading in gpd/sq ft on the trickling filter?

ANS_____

14. A trickling filter, 70 ft in diameter with a media depth of 6 ft, receives a flow of 0.78 MGD. If the BOD concentration of the primary effluent is 167 mg/L, what is the organic loading on the trickling filter in lbs BOD/day/1000 cu ft?

ANS_____

15. The influent to the trickling filter is 1.61 MGD. If the recirculated flow is 2.27 MGD, what is the recirculation ratio?

ANS_____

16. The suspended solids concentration entering a trickling filter is 236 mg/L. If the suspended solids concentration of the trickling filter effluent is 33 mg/L, what is the suspended solids removal efficiency of the trickling filter?

ANS_____

11 *Rotating Biological Contactors*

PRACTICE PROBLEMS 11.1: Hydraulic Loading Rate

1. A rotating biological contactor (RBC) treats a primary effluent flow of 2.94 MGD. If the media surface area is 700,000 sq ft, what is the hydraulic loading rate in gpd/sq ft on the RBC? (Round to the nearest tenth.)

ANS_____

2. A rotating biological contactor treats a flow of 4,654,000 gpd. The manufacturer data indicates a media surface area of 850,000 sq ft. What is the hydraulic loading rate in gpd/sq ft on the RBC? (Round to the nearest tenth.)

ANS_____

3. The manufacturer data indicates a media surface area of 400,000 sq ft. A rotating biological contactor treats a flow of 1.49 MGD. What is the hydraulic loading rate in gpd/sq ft on the RBC? (Round to the nearest tenth.)

ANS_____

4. A rotating biological contactor has a media area of 750,000 sq ft. For a maximum hydraulic loading of 6 gpd/sq ft, what is the desired gpd flow to the contactor?

ANS_____

PRACTICE PROBLEMS 11.2: Soluble BOD

1. The suspended solids concentration of a wastewater is 238 mg/L. If the normal K-value at the plant is 0.45, what is the estimated particulate BOD concentration of the wastewater?

ANS_____

2. The wastewater entering a rotating biological contactor has a BOD content of 218 mg/L. The suspended solids content is 232 mg/L. If the K-value is 0.5, what is the estimated soluble BOD (mg/L) of the wastewater?

ANS_____

3. The wastewater entering a rotating biological contactor has a BOD content of 235 mg/L. The suspended solids concentration of the wastewater is 140 mg/L. If the K-value is 0.5, what is the estimated soluble BOD (mg/L) of the wastewater?

ANS_____

4. A rotating biological contactor receives a 1.8 MGD flow with a BOD concentration of 275 mg/*L* and a SS concentration of 264 mg/*L*. If the K-value is 0.6, how many lbs/day soluble BOD enter the RBC?

ANS_____

PRACTICE PROBLEMS 11.3: Organic Loading Rate

1. A rotating biological contactor (RBC) has a media surface area of 900,000 sq ft and receives a flow of 4,270,000 gpd. If the soluble BOD concentration of the primary effluent is 155 mg/L, what is the organic loading rate in lbs/day/1000 sq ft on the RBC? (Round to the nearest tenth.)

ANS_____

2. A rotating biological contactor (RBC) has a media surface area of 600,000 sq ft and receives a flow of 1,450,000 gpd. If the soluble BOD concentration of the primary effluent is 174 mg/L, what is the organic loading in lbs/day/1000 sq ft on the RBC? (Round to the nearest tenth.)

ANS_____

3. The wastewater flow to an RBC is 2,640,000 gpd. The wastewater has a soluble BOD concentration of 124 mg/L. The RBC media has a total surface area of 650,000 sq ft. What is the organic loading rate in lbs/day/1000 sq ft on the RBC? (Round to the nearest tenth.)

ANS_____

4. A rotating biological contactor (RBC) receives a flow of 2.6 MGD. The BOD of the influent wastewater to the RBC is 182 mg/*L* and the surface area of the media is 725,000 sq ft. If the suspended solids concentration of the wastewater is 135 mg/*L* and the K-value is 0.45, what is the organic loading rate in lbs/day/1000 sq ft? (Round to the nearest tenth.)

ANS_____

Chapter 11—Achievement Test

1. A rotating biological contactor (RBC) treats a primary effluent flow of 2.82 MGD. If the media surface area is 650,000 sq ft, what is the hydraulic loading rate in gpd/sq ft on the RBC? (Round to the nearest tenth.)

ANS_____

2. The suspended solids concentration of a wastewater is 216 mg/L. If the normal K-value at the plant is 0.5, what is the estimated particulate BOD concentration of the wastewater?

ANS_____

3. A rotating biological contactor (RBC) has a media surface area of 700,000 sq ft and receives a flow of 1,860,000 gpd. If the soluble BOD concentration of the primary effluent is 149 mg/L, what is the organic loading in lbs/day/1000 sq ft on the RBC? (Round to the nearest tenth.)

ANS_____

4. A rotating biological contactor receives a 2.7 MGD flow with a BOD concentration of 195 mg/L and SS of 205 mg/L. If the K-value is 0.6, how many lbs/day soluble BOD enter the RBC?

ANS_____

5. A rotating biological contactor treats a flow of 4,276,000 gpd. The manufacturer data indicates a media surface area of 900,000 sq ft. What is the hydraulic loading rate in gpd/sq ft on the RBC? (Round to the nearest tenth.)

ANS_____

6. The wastewater flow to an RBC is 2,316,000 gpd. The wastewater has a soluble BOD concentration of 119 mg/L. The RBC media has a total of 720,000 sq ft. What is the organic loading rate in lbs/day/1000 sq ft on the RBC? (Round to the nearest tenth.)

ANS_____

12 *Activated Sludge*

PRACTICE PROBLEMS 12.1: Aeration Tank, Secondary Clarifier and Oxidation Ditch Volume

1. An aeration tank is 70 ft long, 30 ft wide and operates at an average depth of 15 ft. What is the capacity of the tank, in gallons?

ANS_____

2. What is the gallon capacity of an aeration tank that is 90 ft long, 35 ft wide, and operates at an average depth of 14 ft?

ANS_____

3. A secondary clarifier has a diameter of 70 ft and an average depth of 12 ft. What is the volume of water in the clarifier, in gallons?

ANS_____

4. A clarifier has a diameter of 60 ft and an average depth of 14 ft. What is the volume of water in the clarifier, in gallons?

ANS_____

5. Calculate the cu ft volume of water in the oxidation ditch shown below. The cross section of the ditch is trapezoidal.

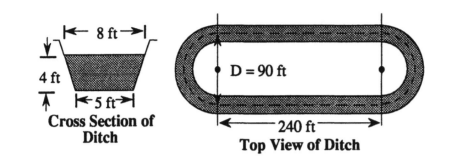

Cross Section of Ditch

Top View of Ditch

ANS_____

6. Calculate the cu ft volume of water in the oxidation ditch shown below. The cross section of the ditch is trapezoidal.

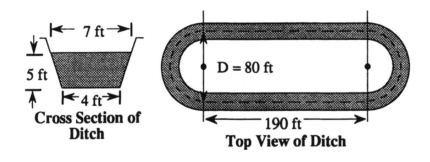

Cross Section of Ditch

Top View of Ditch

ANS_____

PRACTICE PROBLEMS 12.2: BOD or COD Loading , lbs/day

1. The flow to an aeration tank is 890,000 gpd. If the BOD content of the wastewater entering the aeration tank is 230 mg/L, what is the lbs/day BOD loading?

ANS_____

2. The flow to an aeration tank is 2940 gpm. If the COD concentration of the wastewater is 150 mg/L, how many pounds of COD are applied to the aeration tank daily? (Round the flow to the nearest ten thousand.)

ANS_____

3. The BOD content of the wastewater entering an aeration tank is 155 mg/L. If the flow to the aeration tank is 3,120,000 gpd, what is the lbs/day BOD loading?

ANS_____

4. The daily flow to an aeration tank is 4,720,000 gpd. If the COD concentration of the influent wastewater is 140 mg/L, how many pounds of COD are applied to the aeration tank daily?

ANS_____

PRACTICE PROBLEMS 12.3: Solids Inventory in the Aeration Tank

1. If the mixed liquor suspended solids concentration is 2040 mg/L and the aeration tank has a volume of 450,000 gallons, how many pounds of suspended solids are in the aeration tank?

ANS_____

2. The volume of an aeration basin is 180,000 gallons. If the MLVSS concentration is 2160 mg/L, how many pounds of volatile solids are under aeration?

ANS_____

3. The aeration tank of a conventional activated sludge plant has a mixed liquor volatile suspended solids concentration of 2340 mg/L. If the aeration tank is 90 ft long, 40 ft wide, and has wastewater to a depth of 15 ft, how many pounds of MLVSS are under aeration?

ANS_____

4. The aeration tank is 120 ft long, 40 ft wide, and has wastewater to a depth of 15 ft. If the aeration tank of this conventional activated sludge plant has a mixed liquor suspended solids concentration of 2710 mg/L, how many pounds of MLSS are under aeration?

ANS_____

5. An aeration tank is 100 ft long, 45 ft wide, and has wastewater to a depth of 14 ft. If the mixed liquor suspended solids concentration in the aeration tank is 2460 mg/L, with a volatile solids content of 72%, how many pounds of MLVSS are under aeration?

ANS_____

PRACTICE PROBLEMS 12.4: Food/Microorganism Ratio

1. An activated sludge aeration tank receives a primary effluent flow of 2.61 MGD with a BOD concentration of 195 mg/L. The mixed liquor volatile suspended solids concentration is 2560 mg/L and the aeration tank volume is 470,000 gallons. What is the current F/M ratio? (Round to the nearest hundredth.)

ANS_____

2. An activated sludge aeration tank receives a primary effluent flow of 3,260,000 gpd with a BOD of 145 mg/L. The mixed liquor volatile suspended solids is 2490 mg/L and the aeration tank volume is 470,000 gallons. What is the current F/M ratio? (Round to the nearest hundredth.)

ANS_____

3. The flow to a 190,000-gallon oxidation ditch is 310,000 gpd. The BOD concentration of the wastewater is 175 mg/L. If the mixed liquor suspended solids concentration is 2576 mg/L with a volatile solids content of 70%, what is the F/M ratio? (Round to the nearest hundredth.)

ANS_____

4. The desired F/M ratio at an extended aeration activated sludge plant is 0.7 lbs BOD/lb MLVSS. If the 3.1 MGD primary effluent flow has a BOD of 178 mg/L, how many lbs of MLVSS should be maintained in the aeration tank? (Round to the nearest hundredth.)

ANS_____

5. The desired F/M ratio at a particular activated sludge plant is 0.3 lbs BOD/lb MLVSS. If the 2,460,000-gpd primary effluent flow has a BOD concentration of 139 mg/L, how many lbs of MLVSS should be maintained in the aeration tank? (Round to the nearest hundredth.)

ANS_____

PRACTICE PROBLEMS 12.5: Sludge Age (Gould)

1. An aeration tank has a total of 15,600 lbs of mixed liquor suspended solids. If a total of 2520 lbs/day suspended solids enter the aeration tank in the primary effluent flow, what is the sludge age in the aeration tank? (Round to the nearest tenth.)

ANS_____

2. An aeration tank contains 470,000 gallons of wastewater with a MLSS concentration of 2740 mg/L. If the primary effluent flow is 2.8 MGD with a suspended solids concentration of 108 mg/L, what is the sludge age? (Round to the nearest tenth.)

ANS_____

3. An aeration tank is 100 ft long, 40 ft wide, and operates at a depth of 15 ft. The MLSS concentration in the aeration tank is 2480 mg/L. If the influent flow to the tank is 2.72 MGD with a suspended solids concentration of 110 mg/L, what is the sludge age? (Round to the nearest tenth.)

ANS_____

4. The MLSS concentration in the aeration tank is 2940 mg/L. The aeration tank is 100 ft long, 45 ft wide, and operates at a depth of 15 ft. If the influent flow to the tank is 1.94 MGD and has a suspended solids concentration of 105 mg/L, what is the sludge age? (Round to the nearest tenth.)

ANS_____

5. An oxidation ditch has a volume of 210,000 gallons. The 260,000-gpd flow to the oxidation ditch has a suspended solids concentration of 201 mg/L. If the MLSS concentration is 3720 mg/L, what is the sludge age in the oxidation ditch? (Round to the nearest tenth.)

ANS_____

PRACTICE PROBLEMS 12.6: Solids Retention Time

1. An activated sludge system has a total of 28,600 lbs of mixed liquor suspended solids. The suspended solids leaving the final clarifier in the effluent is calculated to be 390 lbs/day. The pounds suspended solids wasted from the final clarifier is 2860 lbs/day. What is the solids retention time, in days?

ANS_____

2. Determine the solids retention time (SRT) given the following data: (Use the "core sampler" method of calculating SRT.)

Aer. Tank Vol.—1,400,000 gal MLSS—2650 mg/*L*
Fin. Clar.—105,000 gal   WAS—5960 mg/*L*
P. E. Flow—3.1 MGD   S.E. SS—20 mg/*L*
WAS Pumping Rate—70,000 gpd CCSS—1920 mg/*L*

ANS_____

3. An aeration tank has a volume of 450,000 gallons. The final clarifier has a volume of 175,000 gallons. The MLSS concentration in the aeration tank is 2115 mg/*L*. If a total of 1570 lbs/day SS are wasted and 230 lbs/day SS are in the secondary effluent, what is the solids retention time for the activated sludge system? (Use the "combined volume" method of calculating system solids.)

ANS_____

4. Calculate the solids retention time (SRT) given the following data:
(Use the "combined volume" method of calculating SRT.)

> Aer. Tank Vol.—340,000 gal
> Fin. Clar.—120,000 gal
> P.E. Flow—1.2 MGD
> WAS Pumping Rate—25,000 gpd

> MLSS—2940 mg/L
> WAS—6110 mg/L
> S.E. SS—17 mg/L

ANS_____

5. The volume of an aeration tank is 365,000 gallons and the final clarifier is 105,000 gallons. The desired SRT for a plant is 7.5 days. The primary effluent flow is 1.76 MGD and the WAS pumping rate is 28,400 gpd. If the WAS SS is 5740 mg/L and the secondary effluent SS is 18 mg/L, what is the desired MLSS mg/L? (Use the "combined volume" method of calculating SRT.)

ANS_____

6. Calculate the solids retention time (SRT) given the following data:
(Use the "core sampler" method of calculating SRT.)

> Aer. Tank Vol.—1.5 MG
> Fin. Clar.—0.11 MG
> P.E. Flow—3.4 MGD
> WAS Pumping Rate—60,000 gpd

> MLSS—2460 mg/L
> WAS—8040 mg/L
> S.E. SS—18 mg/L
> CCSS—1850 mg/L

ANS_____

PRACTICE PROBLEMS 12.7: Return Sludge Rate

1. The settleability test after 30 minutes indicates a sludge settling volume of 215 mL/L. Calculate the RAS flow as a ratio to the secondary influent flow. (Round to the nearest hundredth.)

ANS_____

2. Given the following data, calculate the RAS return rate (Q_{RAS}) using the secondary clarifier solids balance equation.

   MLSS—2460 mg/L Q_{WAS}—60,000 gpd
   RAS SS—7850 mg/L Q_{PE} —3.4 MGD
   WAS SS—7850 mg/L

ANS_____

3. A total of 270 mL/L sludge settled during a settleability test after 30 minutes. The secondary influent flow is 3.12 MGD. (a) Calculate the RAS flow as a ratio to the secondary influent flow. (Round to the nearest hundredth.) (b) What is the RAS flow expressed in MGD? (Round to the nearest hundredth.)

ANS_____

4. The secondary influent flow to an aeration tank is 1.67 MGD. If the results of the settleability test after 30 minutes indicate that 285 mL/L sludge settled. (a) What is the RAS flow as a ratio to the secondary influent flow? (Round to the nearest hundredth.) (b) What is the RAS flow expressed in MGD? (Round to the nearest hundredth.)

ANS_____

5. Given the following data, calculate the RAS return rate using the aeration tank solids balance equation.

   MLSS—2100 mg/L
   RAS SS—7490 mg/L
   Q_{PE}—6.3 MGD

ANS_____

PRACTICE PROBLEMS 12.8: Wasting Rate

1. The desired F/M ratio for an activated sludge system is 0.6 lbs BOD/lb MLVSS. It has been calculated that 3300 lbs of BOD enter the aeration tank daily. If the volatile solids content of the MLSS is 68%, how many lbs MLSS are desired in the aeration tank?

ANS_____

2. Using a desired sludge age, it was calculated that 14,850 lbs MLSS are desired in the aeration tank. If the aeration tank volume is 780,000 gallons and the MLSS concentration is 2670 mg/L, how many lbs/day MLSS should be wasted?

ANS_____

3. Given the following data, determine the lbs/day SS to be wasted:

Aer. Tank Vol.—1,100,000 gal   MLSS—2100 mg/L
Influent Flow—2,930,000 gpd   % VS—67%
BOD—108 mg/L
Desired F/M—0.3 lbs BOD/day/lb MLVSS

ANS_____

4. The desired sludge age for an activated sludge plant is 5.5 days. The aeration tank volume is 850,000 gal. If 3140 lbs/day suspended solids enter the aeration tank and the MLSS concentration is 2920 mg/L, how many lbs/day MLSS (suspended solids) should be wasted?

ANS_____

5. The desired SRT for an activated sludge plant is 8.5 days. There are a total of 32,100 lbs SS in the system. The secondary effluent flow is 3,160,000 gpd with a suspended solids content of 22 mg/L. How many lbs/day WAS SS must be wasted to maintain the desired SRT?

ANS_____

6. Given the following data, calculate the lbs/day WAS SS to be wasted. (Use the "combined volume" method of calculating SRT.)

Desired SRT—10 days   S.E. SS—15 mg/L
Clarifier + Aerator Vol.—1.33 MG   Inf. Flow—6.85 MGD
MLSS—2940 mg/L

ANS_____

PRACTICE PROBLEMS 12.9: WAS Pumping Rate Calculations

1. It has been determined that 5640 lbs/day of dry solids must be removed from the secondary system. If the RAS SS concentration is 6630 mg/L, what must be the WAS pumping rate, in MGD? (Round to the nearest hundredth.)

ANS_____

2. The WAS suspended solids concentration is 6120 mg/L. If 8640 lbs/day dry solids are to be wasted, (a) What must the WAS pumping rate be, in MGD? (Round to the nearest hundredth.) (b) What is this rate expressed in gpm?

ANS_____

3. Given the following data, calculate the WAS pumping rate required (in MGD):
   (Use the "combined volume" method of calculating system solids.)
   (Round the answer to the nearest thousandth.)

   Desired SRT—9.5 days   RAS SS—7410 mg/L
   Clarifier + Aerator Vol.—1.7 MG   S.E. SS—19 mg/L
   MLSS—2630 mg/L   Inf. Flow—4.1 MGD

ANS_____

4. Given the following data, calculate the WAS pumping rate required (in MGD): (Use the "combined volume" method of calculating system solids.) (Round the answer to the nearest thousandth.)

Desired SRT—8 days

Clarifier + Aerator Vol.—1.6 MG

MLSS—2580 mg/L

RAS SS—5990 mg/L

S.E. SS—13 mg/L

Inf. Flow—3.9 MGD

ANS_____

PRACTICE PROBLEMS 12.10: Oxidation Ditch Detention Time

1. An oxidation ditch has a volume of 165,000 gallons. If the flow to the oxidation ditch is 180,000 gpd, what is the detention time, in hours?

ANS_____

2. An oxidation ditch receives a flow of 0.21 MGD. If the volume of the oxidation ditch is 360,000 gallons, what is the detention time in hours?

ANS_____

3. If the volume of the oxidation ditch is 415,000 gallons, and the oxidation ditch receives a flow of 295,000 gpd, what is the detention time in hours?

ANS_____

4. The volume of an oxidation ditch is 190,000 gallons. If the oxidation ditch receives a flow of 305,000 gpd, what is the detention time in hours?

ANS_____

Chapter 12—Achievement Test

1. An aeration tank is 90 ft long, 30 ft wide, and operates at an average depth of 16 ft. What is the capacity of the tank, in gallons?

ANS_____

2. The BOD content of the wastewater entering an aeration tank is 217 mg/L. If the flow to the aeration tank is 1,668,000 gpd, what is the lbs/day BOD loading?

ANS_____

3. The flow to a 210,000-gallon oxidation ditch is 389,000 gpd. The BOD concentration of the wastewater is 218 mg/L. If the mixed liquor suspended solids concentration is 3250 mg/L, with a volatile solids content of 67%, what is the F/M ratio? (Round to the nearest hundredth.)

ANS_____

4. A clarifier has a diameter of 80 ft and an average depth of 10 ft. What is the capacity of the clarifier, in gallons?

ANS_____

5. An activated sludge aeration tank receives a primary effluent flow of 2.13 MGD with a BOD concentration of 175 mg/L. The mixed liquor volatile suspended solids concentration is 2880 mg/L and the aeration tank volume is 420,000 gallons. What is the current F/M ratio? (Round to the nearest hundredth.)

ANS_____

6. Calculate the cu ft capacity of the oxidation ditch shown below. The cross section of the ditch is trapezoidal.

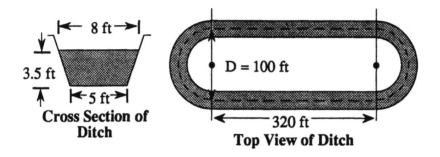

Cross Section of Ditch

Top View of Ditch

ANS_____

7. The daily flow to an aeration tank is 3,840,000 gpd. If the COD concentration of the influent wastewater is 155 mg/L, how many pounds of COD are applied to the aeration tank daily?

ANS_____

8. An aeration tank contains 525,000 gallons of wastewater with a MLSS concentration of 2610 mg/L. If the primary effluent flow is 1.7 MGD with a suspended solids concentration of 185 mg/L, what is the sludge age? (Round to the nearest tenth.)

ANS_____

Chapter 12—Achievement Test—Cont'd

9. The desired F/M ratio at a particular activated sludge plant is 0.4 lbs BOD/lb MLVSS. If the 2.78-MGD primary effluent flow has a BOD concentration of 143 mg/L, how many lbs of MLVSS should be maintained in the aeration tank? (Round to the nearest tenth.)

ANS_____

10. An oxidation ditch receives a flow of 0.28 MGD. If the volume of the oxidation ditch is 390,000 gallons, what is the detention time in hours?

ANS_____

11. The desired F/M ratio at a particular activated sludge plant is 0.7 lbs COD/lb MLVSS. If the 2,390,000-gpd primary effluent flow has a COD concentration of 158 mg/L, how many lbs of MLVSS should be maintained in the aeration tank?

ANS_____

12. An aeration tank is 100 ft long, 45 ft wide, and operates at a depth of 13 ft. The MLSS concentration in the aeration tank is 2830 mg/L. If the influent flow to the tank is 1.1 MGD and contains a suspended solids concentration of 160 mg/L, what is the sludge age? (Round to the nearest tenth.)

ANS_____

13. If the volume of the oxidation ditch is 600,000 gallons, and an oxidation ditch receives a flow of 0.34 MGD, what is the detention time in hours? (Round to the nearest tenth.)

ANS_____

14. An oxidation ditch has a volume of 250,000 gallons. The 0.3-MGD flow to the oxidation ditch has a suspended solids concentration of 195 mg/L. If the MLSS concentration is 3910 mg/L, what is the sludge age in the oxidation ditch? (Round to the nearest tenth.)

ANS_____

15. If the mixed liquor suspended solids concentration is 2660 mg/L, and the aeration tank has a volume of 425,000 gallons, how many pounds of suspended solids are in the aeration tank?

ANS_____

16. The desired F/M ratio at a conventional activated sludge plant is 0.3 lbs BOD/lb MLVSS. If the 2.81-MGD primary effluent flow has a BOD of 144 mg/L, how many lbs of MLVSS should be maintained in the aeration tank?

ANS_____

17. The aeration tank of a conventional activated sludge plant has a mixed liquor volatile suspended solids concentration of 2470 mg/L. If the aeration tank is 100 ft long, 45 ft wide, and has wastewater to a depth of 17 ft, how many pounds of MLVSS are in the aeration tank?

ANS_____

Chapter 12—Achievement Test—Cont'd

18. The MLSS concentration in an aeration tank is 2740 mg/L. The aeration tank contains 705,000 gallons of wastewater. If the primary effluent flow is 1.78 MGD with a suspended solids concentration of 180 mg/L, what is the sludge age? (Round to the nearest tenth.)

ANS_____

19. Determine the solids retention time (SRT) given the following data: (Use the "core sampler" method of calculating system solids.)

Aer. Tank Vol.—1,380,000 gal   MLSS—2650 mg/L
Fin. Clar.—117,000 gal   WAS—5960 mg/L
P.E. Flow—2.9 MGD   S.E. SS—20 mg/L
WAS—75,000 gpd   CCSS—1900 mg/L

ANS_____

20. The settleability test after 30 minutes indicates a sludge settling volume of 228 mL/L. Calculate the RAS flow as a ratio to the secondary influent flow. (Round to the nearest hundredth.)

ANS_____

21. The desired F/M ratio at an activated sludge plant is 0.5 lbs BOD/lb MLVSS. It was calculated that 3630 lbs/day BOD enter the aeration tank. If the volatile solids content of the MLSS is 71%, how many lbs MLSS are desired in the aeration tank?

ANS_____

22. Calculate the solids retention time (SRT) given the following data:
(Use the "combined volume" method of calculating system solids.)

   Aer. Tank Vol.—360,000 gal MLSS—2890 mg/L
   Fin. Clar.—125,000 gal   WAS—6050 mg/L
   P.E. Flow—1.42 MGD   S.E.SS — 22 mg/L
   WAS—28,000 gpd

                 ANS_____

23 The desired sludge age for a plant is 4.8 days. The aeration tank volume is 770,000
gal. If 3670 lbs/day suspended solids enter the aeration tank and the MLSS
concentration is 2730 mg/L, how many lbs/day MLSS (suspended solids) should be
wasted?

                 ANS_____

24. It has been determined that 4100 lbs/day of dry solids must be removed from the
secondary system. If the RAS SS concentration is 6340 mg/L, what must be the WAS
pumping rate, in MGD? (Round to the nearest thousandth.)

                 ANS_____

25. Given the following data, calculate the lbs/day WAS SS to be wasted.
(Use the "combined volume" method of calculating system solids.)

   Desired SRT—10 days   RAS SS—5910 mg/L
   Clarifier + Aerator Vol.—1.45 MG S.E. SS—18 mg/L
   MLSS—2870 mg/L   P.E. Flow—5.68 MGD

                 ANS_____

13 *Waste Treatment Ponds*

PRACTICE PROBLEMS 13.1: BOD Loading

1. Calculate the BOD loading (lbs/day) on a pond if the influent flow is 390,000 gpd with a BOD of 245 mg/L.

ANS_____

2. The BOD concentration of the wastewater entering a pond is 158 mg/L. If the flow to the pond is 220,000 gpd, how many lbs/day BOD enter the pond?

ANS_____

3. The flow to a waste treatment pond is 175 gpm. If the BOD concentration of the water is 221 mg/L, how many pounds of BOD are applied to the pond daily?

ANS_____

4. The BOD concentration of the influent wastewater to a waste treatment pond is 190 mg/*L*. If the flow to the pond is 125 gpm, how many pounds of BOD are applied to the pond daily?

ANS_____

PRACTICE PROBLEMS 13.2: Organic Loading Rate

1. A 7.5-acre pond receives a flow of 200,000 gpd. If the influent flow has a BOD content of 190 mg/*L*, what is the organic loading rate in lbs/day/ac on the pond?

ANS_____

2. A pond has an average width of 400 ft and an average length of 710 ft. The flow to the pond is 157,000 gpd with a BOD content of 147 mg/*L*. What is the organic loading rate in lbs/day/ac on the pond? (Round acres to the nearest tenth.)

ANS_____

3. The flow to a pond is 70,000 gpd with a BOD content of 124 mg/*L*. The pond has an average width of 220 ft and an average length of 382 ft. What is the organic loading rate in lbs/day/ac on the pond? (Round acres to the nearest tenth.)

ANS_____

4. The maximum desired organic loading rate for a 15-acre pond is 20 lbs BOD/day/ac. If the influent flow to the pond has a BOD concentration of 187 mg/L, what is the maximum desirable MGD flow to the pond? (Round to the nearest hundredth.)

ANS_____

PRACTICE PROBLEMS 13.3: BOD Removal Efficiency

1. The BOD entering a waste treatment pond is 207 mg/L. If the BOD in the pond
    effluent is 39 mg/L, what is the BOD removal efficiency of the pond?

    ANS_____

2. The influent of a waste treatment pond has a BOD content 262 mg/L. If the BOD
    content of the pond effluent is 130 mg/L, what is the BOD removal efficiency of the
    pond?

    ANS_____

3. The BOD entering a waste treatment pond is 280 mg/L. If the BOD in the pond
    effluent is 45 mg/L, what is the BOD removal efficiency of the pond?

    ANS_____

4. The BOD entering a waste treatment pond is 140 mg/*L*. If the BOD in the pond effluent is 56 mg/*L*, what is the BOD removal efficiency of the pond?

ANS_____

PRACTICE PROBLEMS 13.4: Hydraulic Loading Rate

1. A 20-acre pond receives a flow of 3.3 acre-feet/day. What is the hydraulic loading rate on the pond in in./day?

ANS_____

2. A 15-acre pond receives a flow of 5 acre-feet/day. What is the hydraulic loading rate on the pond in in./day?

ANS_____

3. A waste treatment pond receives a flow of 2,320,000 gpd. If the surface area of the pond is 18 acres, what is the hydraulic loading in in./day?

ANS_____

4. A waste treatment pond receives a flow of 1,680,000 gpd. If the surface area of the pond is 14 acres, what is the hydraulic loading in in./day?

ANS_____

PRACTICE PROBLEMS 13.5: Population Loading and Population Equivalent

1. A 4-acre wastewater pond serves a population of 1320 people. What is the population loading (people/acre) on the pond?

ANS_____

2. A wastewater pond serves a population of 5460 people. If the pond covers 18.5 acres, what is the population loading (people/acre) on the pond?

ANS_____

3. A 0.6-MGD wastewater flow has a BOD concentration of 1680 mg/L. Using an average of 0.2 lbs BOD/day/person, what is the population equivalent of this wastewater flow?

ANS_____

4. A 250,000-gpd wastewater flow has a BOD content of 2160 mg/*L*. Using an average of 0.2 lbs BOD/day/person, what is the population equivalent of this flow?

ANS_____

PRACTICE PROBLEMS 13.6: Detention Time

1. A waste treatment pond has a total volume of 17 ac-ft. If the flow to the pond is 0.42 ac-ft/day, what is the detention time of the pond (days)?

ANS_____

2. A waste treatment pond is operated at a depth of 6 feet. The average width of the pond is 440 ft and the average length is 680 ft. If the flow to the pond is 0.3 MGD, what is the detention time, in days?

ANS_____

3. The average width of the pond is 240 ft and the average length is 390 ft. A waste treatment pond is operated at a depth of 5 feet. If the flow to the pond is 70,000 gpd, what is the detention time, in days?

ANS_____

4. A waste treatment pond has an average length of 680 ft, an average width of 420 ft, and a water depth of 4 ft. If the flow to the pond is 0.47 ac-ft/day what is the detention time for the pond?

ANS_____

Chapter 13—Achievement Test

1. The BOD concentration of the wastewater entering a pond is 190 mg/L. If the flow to the pond is 360,000 gpd, how many lbs/day BOD enter the pond?

ANS_____

2. A 8.5-acre pond receives a flow of 280,000 gpd. If the influent flow has a BOD content of 240 mg/L, what is the organic loading rate in lbs/day/ac on the pond?

ANS_____

3. The BOD entering a waste treatment pond is 210 mg/L. If the BOD concentration in the pond effluent is 43 mg/L, what is the BOD removal efficiency of the pond?

ANS_____

4. A 22-acre pond receives a flow of 3.6 acre-feet/day. What is the hydraulic loading on the pond in in./day?

ANS_____

5. The BOD entering a waste treatment pond is 162 mg/L. If the BOD concentration in the pond effluent is 71 mg/L, what is the BOD removal efficiency of the pond?

ANS_____

6. The flow to a waste treatment pond is 200 gpm. If the BOD concentration of the water is 218 mg/L, how many pounds of BOD are applied to the pond daily?

ANS_____

Chapter 13—Achievement Test—Cont'd

7. The flow to a pond is 75,000 gpd with a BOD content of 138 mg/*L*. The pond has an average width of 230 ft and an average length of 390 ft. What is the organic loading rate in lbs/day/ac on the pond?

ANS_____

8. A waste treatment pond receives a flow of 1,960,000 gpd. If the surface area of the pond is 19 acres, what is the hydraulic loading in in./day?

ANS_____

9. A wastewater pond serves a population of 6000 people. If the area of the pond is 20 acres, what is the population loading on the pond?

ANS_____

10. A waste treatment pond has a total volume of 18.2 ac-ft. If the flow to the pond is 0.51 ac-ft/day, what is the detention time of the pond, in days?

ANS_____

11. A 0.7-MGD wastewater flow has a BOD concentration of 2840 mg/L. Using an average of 0.2 lbs/day BOD/person, what is the population equivalent of this wastewater flow?

ANS_____

12. A waste treatment pond is operated at a depth of 5 feet. The average width of the pond is 420 ft and the average length is 710 ft. If the flow to the pond is 0.35 MGD, what is the detention time, in days?

ANS_____

14 Chemical Dosage

PRACTICE PROBLEMS 14.1: Chemical Feed Rate—
(Full-Strength Chemicals)

1. Determine the chlorinator setting (lbs/day) needed to treat a flow of 4.4 MGD with a chlorine dose of 3.2 mg/L.

ANS_____

2. The desired dosage for a dry polymer is 10 mg/L. If the flow to be treated is 1,660,000 gpd, how many lbs/day polymer will be required?

ANS_____

3. To neutralize a sour digester, one pound of lime is to be added for every pound of volatile acids in the digester sludge. If the digester contains 195,000 gal of sludge with a volatile acid (VA) level of 2100 mg/L, how many pounds of lime should be added?

ANS_____

4. A total of 312 lbs of chlorine was used during a 24-hr period to chlorinate a flow of 4.88 MGD. At this lbs/day dosage rate, what was the mg/L dosage rate?

ANS_____

PRACTICE PROBLEMS 14.2: Chlorine Dose, Demand, and Residual

1. The secondary effluent is tested and found to have a chlorine demand of 4.8 mg/L. If the desired chlorine residual is 0.9 mg/L, what is the desired chlorine dose in mg/L?

ANS_____

2. The chlorine dosage for a secondary effluent is 8.4 mg/L. If the chlorine residual after 30 minutes contact time is found to be 0.8 mg/L, what is the chlorine demand expressed in mg/L?

ANS_____

3. The chlorine demand of a secondary effluent is 7.7 mg/L. If a chlorine residual of 0.5 mg/L is desired, what is the desired chlorine dosage in mg/L?

ANS_____

4. What should the chlorinator setting be (lbs/day) to treat a flow of 3.9 MGD if the chlorine demand is 8 mg/*L* and a chlorine residual of 1.5 mg/*L* is desired?

ANS_____

PRACTICE PROBLEMS 14.3: Chemical Feed Rate—
(Chemicals Less Than Full Strength)

1. A total chlorine dosage of 10.8 mg/L is required to treat a particular water. If the flow is 2.77 MGD and the hypochlorite has 65% available chlorine, how many lbs/day of hypochlorite will be required?

ANS_____

2. The desired dose of a polymer is 9.5 mg/L. The polymer literature indicates that the polymer compound provided is 60% active polymer. If a flow of 3.52 MGD is to be treated, how many lbs/day of the polymer compound will be required?

ANS_____

3. A wastewater flow of 1,695,000 gpd requires a chlorine dose of 18 mg/L. If hypochlorite (65% available chlorine) is to be used, how many lbs/day of hypochlorite are required?

ANS_____

4. A total of 900 lbs of 65% hypochlorite are used in a day. If the flow rate treated is 5.15 MGD, what is the chlorine dosage in mg/L?

ANS_____

PRACTICE PROBLEMS 14.4: Percent Strength of Solutions

1. If a total of 10 ounces of dry polymer are added to 15 gallons of water, what is the percent strength (by weight) of the polymer solution? (Round the pounds polymer to the nearest tenth and round the answer to the nearest tenth.)

ANS_____

2. How many pounds of dry polymer must be added to 20 gallons of water to make a 0.8% polymer solution? (Round to the nearest tenth.)

ANS_____

3. If 150 grams of dry polymer are dissolved in 10 gallons of water, what percent strength is the solution? (Round to the nearest tenth.) (1 g = 0.0022 lbs)

ANS_____

4. A 10% liquid polymer is to be used in making up a polymer solution. How many lbs of liquid polymer should be mixed with water to produce 167 lbs of a 0.5% polymer solution?

ANS_____

5. A 10% liquid polymer will be used in making up a solution. How many gallons of liquid polymer should be added to the water to make up 50 gallons of a 0.3% polymer solution? The liquid polymer has a specific gravity of 1.25, and assume the polymer solution has a specific gravity of 1.0. (Round to the nearest tenth.)

ANS_____

6. How many gallons of 12% liquid polymer should be mixed with water to produce 100 gallons of a 0.6% polymer solution? The density of the polymer liquid is 10.3 lbs/gal. Assume the density of the polymer solution is 8.34 lbs/gal.

ANS_____

PRACTICE PROBLEMS 14.5: Mixing Solutions of Different Strength

1. If 25 lbs of a 10% strength solution are mixed with 100 lbs of a 0.5% strength solution, what is the percent strength of the solution mixture? (Round to the nearest tenth.)

ANS_____

2. If 5 gallons of a 12% strength solution are added to 32 gallons of a 0.3% strength solution, what is the percent strength of the solution mixture? Assume the 12% strength solution weighs 10.4 lbs/gal and the 0.3% strength solution weighs 8.4 lbs/gal. (Round to the nearest tenth.)

ANS_____

3. If 10 gallons of a 10% strength solution is mixed with 40 gallons of a 0.25% strength solution, what is the percent strength of the solution mixture? Assume the 10% solution weighs 10.2 lbs/gal and the 0.25% solution weighs 8.34 lbs/gal. (Round to the nearest tenth.)

ANS_____

4. What weights of a 1% solution and a 6% solution must be mixed to make 500 lbs of a 5% solution?

ANS_____

5. How many lbs of a 1% polymer solution and water should be mixed together to form 100 lbs of a 0.4% polymer solution?

ANS_____

6. What weights of a 0.5% solution and a 10% solution must be mixed to make 250 lbs of a 2% solution?

ANS_____

PRACTICE PROBLEMS 14.6: Solution Chemical Feeder Setting, gpd

1. Jar tests indicate that the best liquid alum dose for a water is 11 mg/L. The flow to be treated is 3.89 MGD. Determine the gpd setting for the liquid alum chemical feeder if the liquid alum contains 5.36 lbs of alum per gallon of solution.

ANS_____

2. Jar tests indicate that the best liquid alum dose for a water is 9 mg/L. The flow to be treated is 1,340,000 gpd. Determine the gpd setting for the liquid alum chemical feeder if the liquid alum contains 5.36 lbs of alum per gallon of solution.

ANS_____

3. Jar tests indicate that the best liquid alum dose for a water is 10 mg/L. The flow to be treated is 2.02 MGD. Determine the gpd setting for the liquid alum chemical feeder if the liquid alum is a 60% solution. (Assume the alum solution weighs 8.34 lbs/gal.)

ANS_____

4. The flow to a plant is 4,120,000 gpd. Jar testing indicates that the optimum alum dose is 8 mg/*L*. What should the gpd setting be for the solution feeder if the alum solution is a 60% solution? Assume the solution weighs 8.34 lbs/gal. (Round to the nearest tenth.)

ANS_____

PRACTICE PROBLEMS 14.7: Chemical Feed Pump—
Percent Stroke Setting

1. The required chemical pumping rate has been calculated to be 25 gpm. If the maximum pumping rate is 90 gpm, what should the percent stroke setting be?

ANS_____

2. The required chemical pumping rate has been calculated to be 20 gpm. If the maximum pumping rate is 85 gpm, what should the percent stroke setting be?

ANS_____

3. The required chemical pumping rate has been determined to be 15 gpm. What is the percent stroke setting if the maximum rate is 80 gpm?

ANS_____

4. The maximum pumping rate is 100 gpm. If the required pumping rate is 30 gpm, what is the percent stroke setting?

ANS_____

PRACTICE PROBLEMS 14.8: Solution Chemical Feeder Setting, m*L*/min

1. The desired solution feed rate was calculated to be 40 gpd. What is this feed rate expressed as m*L*/min?

ANS_____

2. The desired solution feed rate was calculated to be 35 gpd. What is this feed rate expressed as m*L*/min?

ANS_____

3. The optimum polymer dose has been determined to be 8 mg/*L*. The flow to be treated is 890,000 gpd. If the solution to be used contains 60% active polymer, what should the solution chemical feeder setting be, in m*L*/min? Assume the polymer solution weighs 8.34 lbs/gal.

ANS_____

4. The flow to be treated is 1,240,000 gpd. The optimum polymer dose has been determined to be 10 mg/L. If the solution to be used contains 55% active polymer, what should the solution chemical feeder setting be, in mL/min? Assume the polymer solution weighs 8.34 lbs/gal.

ANS_____

PRACTICE PROBLEMS 14.9: Dry Chemical Feeder Calibration

1. Calculate the actual chemical feed rate, in lbs/day, if a bucket is placed under a chemical feeder and a total of 1.8 lbs is collected during a 30-minute period.

ANS_____

2. Calculate the actual chemical feed rate, in lbs/day, if a bucket is placed under a chemical feeder and a total of 1 lb 6 oz is collected during a 25-minute period.

ANS_____

3. To calibrate a chemical feeder, a bucket is first weighed (10 oz) then placed under the chemical feeder. After 30 minutes the bucket is weighed again. If the weight of the bucket with chemical is 1.95 lbs, what is the actual chemical feed rate, in lbs/day?

ANS_____

4. A chemical feeder is to be calibrated. The bucket to be used to collect chemical is placed under the chemical feeder and weighed (0.4 lbs). After 30 minutes, the weight of the bucket and chemical is found to be 2.3 lbs. Based on this test, what is the actual chemical feed rate, in lbs/day?

ANS_____

PRACTICE PROBLEMS 14.10: Solution Chemical Feeder Calibration (Given mL/min Flow)

1. A calibration test is conducted for a solution chemical feeder. During 5 minutes, a total of 720 mL is delivered by the solution feeder. The polymer solution is a 1.3% solution. What is the lbs/day polymer feed rate? Assume the polymer solution weighs 8.34 lbs/gal. (Round to the nearest tenth.)

ANS_____

2. A calibration test is conducted for a solution chemical feeder. During the 5-minute test, the pump delivered 950 mL of the 1.4% polymer solution. What is the polymer dosage rate in lbs/day? Assume the polymer solution weighs 8.34 lbs/gal. (Round to the nearest tenth.)

ANS_____

3. A calibration test is conducted for a solution chemical feeder. During a 5-minute test, the pump delivered 600 mL of a 1.1% polymer solution. The specific gravity of the polymer solution is 1.1. What is the polymer dosage rate in lbs/day? (Round to the nearest tenth.)

ANS_____

4. During a 5-minute test, the pump delivered 810 m*L* of a 0.5% polymer solution. A calibration test is conducted for the solution chemical feeder. The specific gravity of the polymer solution is 1.04. What is the polymer dosage rate in lbs/day? (Round to the nearest tenth.)

ANS_____

PRACTICE PROBLEMS 14.11: Solution Chemical Feeder Calibration (Given Drop in Solution Tank Level)

1. A pumping rate calibration test is conducted for a 3-minute period. The liquid level in the 3-ft diameter solution tank is measured before and after the test. If the level drops 1.3 ft during the 3-minute test, what is the pumping rate in gpm?

ANS_____

2. A pumping rate calibration test is conducted for a 5-minute period. The liquid level in the 4 ft diameter solution tank is measured before and after the test. If the level drops 14 inches during the test, what is the pumping rate in gpm?

ANS_____

3. A pump test indicates that a pump delivers 25 gpm during a 4-minute pumping test. The diameter of the solution tank is 5 feet. What was the expected drop, in ft, in solution level during the pumping test?

ANS_____

4. The liquid level in the 4-ft diameter solution tank is measured before and after the test. A pumping rate calibration test is conducted for a 3-minute period. If the level drops 17 inches during the test, what is the pumping rate in gpm?

ANS_____

PRACTICE PROBLEMS 14.12: Average Use Calculations

1. The amount of chemical used for each day during a week is given below. Based on this data, what was the average lbs/day chemical use during the week?

   Monday—78 lbs/day   Friday—77 lbs/day
   Tuesday—71 lbs/day   Saturday—82 lbs/day
   Wednesday—72 lbs/day   Sunday—85 lbs/day
   Thursday—65 lbs/day

   ANS_____

2. The average chemical use at a plant is 95 lbs/day. If the chemical inventory in stock is 2100 lbs, how many days' supply is this? (Round to the nearest tenth.)

   ANS_____

3. The chemical inventory in stock is 921 lbs. If the average chemical use at a plant is 60 lbs/day, how many days' supply is this? (Round to the nearest tenth.)

   ANS_____

4. The average gallons of polymer solution used each day at a treatment plant is 92 gpd. A chemical feed tank has a diameter of 4 ft and contains solution to a depth of 3 ft 2 inches. How many days' supply are represented by the solution in the tank? (Round to the nearest tenth.)

ANS_____

Chapter 14—Achievement Test

1. The desired dosage for a dry polymer is 9 mg/L. If the flow to be treated is 3,175,000 gpd, how many lbs/day polymer will be required?

ANS_____

2. A total chlorine dosage of 6.9 mg/L is required to treat a particular water. If the flow is 3.12 MGD and the hypochlorite has 65% available chlorine how many lbs/day of hypochlorite will be required?

ANS_____

3. How many pounds of dry polymer must be added to 30 gallons of water to make a 0.1% polymer solution?

ANS_____

4. How many lbs of an 8% polymer solution and water should be mixed together to form 100 lbs of a 1.5% polymer solution?

ANS_____

5. Calculate the actual chemical feed rate, in lbs/day, if a bucket is placed under a chemical feeder and a total of 1.7 lbs is collected during a 30-minute period.

ANS_____

6. Jar tests indicate that the best liquid alum dose for a water is 11 mg/L. The flow to be treated is 2,640,000 gpd. Determine the gpd setting for the liquid alum chemical feeder if the liquid alum contains 5.36 lbs of alum per gallon of solution.

ANS_____

7. A total of 372 lbs of chlorine was used during a 24-hr period to chlorinate a flow of 4,924,000 gpd. At this lbs/day dosage rate, what was the mg/L dosage rate?

ANS_____

8. To calibrate a chemical feeder, a bucket is first weighed (14 oz) then placed under the chemical feeder. After 30 minutes the bucket is weighed again. If the weight of the bucket with the chemical is 2 lbs 3 oz, what is the actual chemical feed rate, in lbs/day?

ANS_____

Chapter 14—Achievement Test—Cont'd

9. The flow to a plant is 3,221,000 gpd. Jar testing indicates that the optimum alum dose is 9 mg/L. What should the gpd setting be for the solution feeder if the alum solution is a 60% solution? Assume the alum solution weighs 8.34 lbs/gal. (Round to the nearest tenth.)

ANS_____

10. The desired chemical pumping rate has been calculated as 25 gpm. If the maximum pumping rate is 80 gpm, what should the percent stroke setting be?

ANS_____

11. The chlorine dosage for a secondary effluent is 7.6 mg/L. If the chlorine residual after 30 minutes contact time is found to be 0.4 mg/L, what is the chlorine demand expressed in mg/L?

ANS_____

12. How many gallons of a 10% liquid polymer should be mixed with water to produce 50 gallons of a 0.5% polymer solution? The density of the polymer liquid is 9.8 lbs/gal. Assume the density of the polymer solution is 8.34 lbs/gal. (Round to the nearest tenth.)

ANS_____

13. A calibration test is conducted for a solution chemical feeder. During 5 minutes, a total of 640 m*L* is delivered by the solution feeder. The polymer solution is a 1.1% solution. What is the lbs/day feed rate? Assume the polymer solution weighs 8.34 lbs/gal. (Round to the nearest tenth.)

ANS_____

14. A pump operates at a rate of 25 gpm. How many feet will the liquid level be expected to drop after a 5-minute pumping test if the diameter of the solution tank is 5 ft? (Round to the nearest tenth.)

ANS_____

15. The desired chemical pumping rate has been calculated to be 15 gpm. If the maximum pumping rate is 80 gpm, what should the percent stroke setting be?

ANS_____

16. What should the chlorinator setting be (lbs/day) to treat a flow of 4.1 MGD if the chlorine demand is 8.6 mg/*L* and a chlorine residual of 0.8 mg/*L* is desired?

ANS_____

Chapter 14—Achievement Test—Cont'd

17. The average chemical use at a plant is 80 lbs/day. If the chemical inventory in stock is 1960 lbs, how many days' supply is this? (Round to the nearest tenth.)

ANS_____

18. The desired solution feed rate was calculated to be 40 gpd. What is this feed rate expressed as mL/min?

ANS_____

19. A calibration test is conducted for a solution chemical feeder. During a 5-minute test, the pump delivered 810 mL of the 0.8% polymer solution. What is the polymer dosage rate in lbs/day? Assume the polymer solution weighs 8.34 lbs/gal. (Round to the nearest tenth.)

ANS_____

20. The flow to be treated is 3,142,000 gpd. The optimum polymer dose has been determined to be 8 mg/L. If the solution to be used contains 55% active polymer, what should the solution chemical feeder setting be, in mL/min?

ANS_____

21. A pumping calibration test is conducted for a 3-minute period. The liquid level in the 3-ft diameter solution tank is measured before and after the test. If the level drops 17 inches during the 3-minute test, what is the pumping rate in gpm?

ANS_____

22. A wastewater flow of 2,942,000 gpd requires a chlorine dose of 10.4 mg/L. If hypochlorite (65% available chlorine) is to be used, how many lbs/day of hypochlorite are required?

ANS_____

23. What weights of a 0.5% solution and a 15% solution must be mixed to make 100 lbs of a 1% solution?

ANS_____

24. If 5 gallons of a 10% strength solution are mixed with 20 gallons of a 0.3% strength solution, what is the percent strength of the solution mixture? Assume the 10% solution weighs 10.1 lbs/gal and the 0.3% solution weighs 8.34 lbs/gal. (Round to the nearest tenth.)

ANS_____

15 *Sludge Production and and Thickening*

PRACTICE PROBLEMS 15.1: Primary and Secondary Clarifier Solids Production

1. A primary clarifier receives a flow of 4.73 MGD with a suspended solids concentration of 282 mg/L. If the clarifier effluent has a suspended solids concentration of 127 mg/L, how many pound of dry solids are generated daily?

ANS_____

2. The suspended solids concentration of the primary influent is 305 mg/L and that of the primary effluent is 128 mg/L. How many pounds of dry solids are produced if the flow is 3.6 MGD?

ANS_____

3. The 1.9-MGD influent to the secondary system has a BOD concentration of 240 mg/L. The secondary effluent contains 122 mg/L BOD. If the bacteria growth rate, Y-value, for this plant is 0.5 lbs SS/lb BOD removed, how many pounds of dry solids are produced each day by the secondary system?

ANS_____

4. The *Y*-value for a treatment plant secondary system is 0.45 lbs SS/lb BOD removed. The influent to the secondary system is 2.67 MGD. If the BOD concentration of the secondary influent is 283 mg/L and the effluent BOD is 125 mg/L, how many pounds of dry solids are produced each day by the secondary system?

ANS_____

PRACTICE PROBLEMS 15.2: Percent Solids and Sludge Pumping

1. The total weight of a sludge sample is 28 grams (sludge sample only, not the dish). If the weight of the solids after drying is 0.56 grams, what is the percent total solids of the sludge?

ANS_____

2. A total of 8,300 lbs/day SS are removed from a primary clarifier and pumped to a sludge thickener. If the sludge has a solids content of 4.1%, how many lbs/day sludge are pumped to the thickener?

ANS_____

3. A total of 9200 gallons of sludge is pumped to a digester. If the sludge has a 5.3% solids content, how many lbs/day solids are pumped to the digester? (Assume the sludge weighs 8.34 lbs/gal.)

ANS_____

4. It is anticipated that 1,480 lbs/day SS will be pumped from the primary clarifier of a new plant. If the primary clarifier sludge has a solids content of 5.1%, how many gpd sludge will be pumped from the clarifier? (Assume the sludge weighs 8.34 lbs/gal.)

ANS_____

5. A primary sludge has a solids content of 4.6%. If 920 lbs/day suspended solids are pumped from the primary clarifier, how many gpd sludge will be pumped from the clarifier? (Assume the sludge weighs 8.34 lbs/gal).

ANS_____

PRACTICE PROBLEMS 15.3: Sludge Thickening

1. A total of 21,500 lbs/day sludge is pumped to a thickener. The sludge has a 4% solids content. If the sludge is concentrated to 6.5% solids, what will be the expected lbs/day sludge flow from the thickener?

ANS_____

2. A primary clarifier sludge has a 5% solids content. If 2850 gpd of primary sludge is pumped to a thickener and the thickened sludge has a solids content of 7%, what would be the expected gpd flow of thickened sludge? (Assume the primary sludge weighs 8.34 lbs/gal and the thickened sludge weighs 8.61 lbs/gal.)

ANS_____

3. A primary clarifier sludge has a 3.5% solids content. A total of 12,600 lbs/day sludge is pumped to a thickener. If the sludge has been concentrated to 5.5% solids, what will be the expected lbs/day sludge flow from the thickener?

ANS_____

4. The sludge from a primary clarifier has a solids content of 3.8%. The primary sludge is pumped at a rate of 5600 gpd to a thickener. If the thickened sludge has a solids content of 6.3%, what is the anticipated gpd sludge flow through the thickener? (Assume the primary sludge weighs 8.34 lbs/gal and the secondary sludge weighs 8.58 lbs/gal.)

ANS_____

PRACTICE PROBLEMS 15.4: Gravity Thickening

1. A gravity thickener 25 ft in diameter receives a flow of 60 gpm primary sludge combined with an 85 gpm secondary effluent flow. What is the hydraulic loading on the thickener in gpd/sq ft?

ANS_____

2. The primary sludge flow to a gravity thickener is 95 gpm. This is blended with a 70 gpm secondary effluent flow. If the thickener has a diameter of 25 ft, what is the hydraulic loading rate in gpd/sq ft?

ANS_____

3. A primary sludge flow equivalent to 110,000 gpd is pumped to a 40-ft diameter gravity thickener. If the solids concentration of the sludge is 3.8%, what is the solids loading in lbs/day/sq ft?

ANS_____

4. What is the solids loading on a gravity thickener (in lbs/day/sq ft), if the primary sludge flow to the 30-ft diameter gravity thickener is 50 gpm, with a solids concentration of 3.6%?

ANS_____

5. A gravity thickener 45 ft in diameter has a sludge blanket depth of 3.5 ft. If sludge is pumped from the bottom of the thickener at the rate of 25 gpm, what is the sludge detention time (in days) in the thickener?

ANS_____

6. A gravity thickener 35 ft in diameter has a sludge blanket depth of 4.1 ft. If the sludge is pumped from the bottom of the thickener at a rate of 28 gpm, what is the sludge detention time, in hours, in the thickener?

ANS_____

7. What is the efficiency of the gravity thickener if the influent flow to the thickener has a sludge solids concentration of 3% and the effluent flow has a sludge solids concentration of 0.8%? Calculate the efficiency using mg/L solids concentration data.

ANS_____

8. The sludge flow entering a gravity thickener contains 3.3 % sludge solids. The effluent from the thickener contains 0.7% sludge solids. What is the efficiency of the gravity thickener in removing sludge solids? Calculate the efficiency using percent solids data.

ANS_____

PRACTICE PROBLEMS 15.4: Gravity Thickening—Cont'd

9. The sludge solids concentration of the influent flow to a gravity thickener is 3.2%. If the sludge withdrawn from the bottom of the thickener has a sludge solids concentration of 8.1%, what is the concentration factor?

ANS_____

10. The influent flow to a gravity thickener has a sludge solids concentration of 2.9%. What is the concentration factor if the sludge solids concentration of the sludge withdrawn from the thickener is 7.8%.

ANS_____

11. Given the data below, determine whether the sludge blanket in the gravity thickener is expected to increase, decrease, or remain the same.

    Sludge Pumped to Thickener—120 gpm
    Thickener Sludge Pumped from Thickener—45 gpm
    Primary Sludge Solids—3.2%
    Thickened Sludge Solids—7.8%
    Thickener Effluent Suspended Solids—580 mg/L

ANS_____

12. Given the data below, (a) determine whether the sludge blanket in the gravity thickener is expected to increase, decrease, or remain the same; and (b) if there is an increase or decrease, how many lbs/day should this change be?

    Sludge Pumped to Thickener—105 gpm
    Thickener Sludge Pumped from Thickener—55 gpm
    Primary Sludge Solids—3.4%
    Thickened Sludge Solids—6.9%
    Thickener Effluent Suspended Solids—510 mg/L

ANS_____

13. If solids are being stored at a rate of 9000 lbs/day in a 25-ft diameter gravity thickener, how many hours will it take the sludge blanket to rise 1.5 ft. The solids concentration of the thickened sludge is 6.5%? (Round to the nearest tenth.)

ANS_____

14. Solids are being stored at a rate of 12,000 lbs/day in a 25-ft diameter gravity thickener. How many hours will it take the sludge blanket to rise 2.3 ft? The solids concentration of the thickened sludge is 7%. (Round to the nearest tenth.)

ANS_____

15. After several hours of startup of a gravity thickener, the sludge blanket level is measured at 2.8 ft. The desired sludge blanket level is 5 ft. If the sludge solids are entering the thickener at a rate of 50 lbs/min, what is the desired sludge withdrawal rate, in gpm? The thickened sludge solids concentration is 5.8%

ANS_____

16. The sludge blanket level is measured at 3.1 ft after several hours of startup of a gravity thickener. If the desired sludge blanket level is 6 ft and the sludge solids are entering the thickener at a rate of 58 lbs/min, what is the desired sludge withdrawal rate, in gpm? The thickened sludge solids concentration is 5.5%

ANS_____

PRACTICE PROBLEMS 15.5: Dissolved Air Flotation Thickening

1. A dissolved air flotation (DAF) thickener receives a sludge flow of 880 gpm. If the DAF unit is 45 ft in diameter, what is the hydraulic loading rate, in gpm/sq ft? (Round to the nearest tenth.)

ANS_____

2. A dissolved air flotation (DAF) thickener 30 ft in diameter receives a sludge flow of 640 gpm. What is the hydraulic loading rate, in gpm/sq ft? (Round to the nearest tenth.)

ANS_____

3. The sludge flow to a 35-ft diameter dissolved air thickener is 150,000 gpd. If the influent waste activated sludge has a suspended solids concentration of 8300, what is the solids loading rate in lbs/hr/sq ft? Assume the sludge weighs 8.34 lbs/gal. (Round to the nearest tenth.)

ANS_____

4. The sludge flow to a dissolved air flotation thickener is 125 gpm with a suspended solids concentration of 0.8%. If the DAF unit is 60 ft long and 15 ft wide, what is the solids loading rate in lbs/hr/sq ft? Assume the sludge weighs 8.34 lbs/gal. (Round to the nearest tenth.)

ANS_____

5. The air rotameter indicates 8 cfm is supplied to the dissolved air flotation thickener. What is this air supply expressed as lbs/hr?

ANS_____

6. The air rotameter for the dissolved air flotation thickener indicates 10 cfm is supplied to the DAF unit. What is this air supply expressed as lbs/hr?

ANS_____

7. A dissolved air flotation thickener receives an 80-gpm flow of waste activated sludge with a solids concentration of 8400 mg/L. If air is supplied at a rate of 6 cfm, what is the air-to-solids ratio? (Round to the nearest hundredth.)

ANS_____

8. The sludge flow to a dissolved air flotation thickener is 50 gpm with a suspended solids concentration of 7700 mg/L. If the air supplied to the DAF unit is 4 cfm, what is the air-to-solids ratio? (Round to the nearest hundredth.)

ANS_____

PRACTICE PROBLEMS 15.5: Dissolved Air Flotation Thickening—Cont'd

9. A dissolved air flotation thickener receives a sludge flow of 80 gpm. If the recycle rate is 90 gpm, what is the percent recycle rate?

ANS_____

10. The desired percent recycle rate for a dissolved air flotation unit is 110%. If the sludge flow to the thickener is 60 gpm, what should the recycle flow be, in MGD?

ANS_____

11. A 75-ft diameter DAF thickener receives a sludge flow with a solids concentration of 7600 mg/L. If the effluent solids concentration is 230 mg/L, what is the solids removal efficiency?

ANS_____

12. The solids concentration of the influent sludge to a dissolved air flotation unit is 8250 mg/L. If the thickened sludge solids concentration is 4.6%, what is the concentration factor?

ANS_____

PRACTICE PROBLEMS 15.6: Centrifuge Thickening

1. A disc centrifuge receives a waste activated sludge flow of 35 gpm. What is the hydraulic loading on the unit, in gal/hr?

ANS_____

2. The waste activated sludge flow to a scroll centrifuge thickener is 85,400 gpd. What is the hydraulic loading on the thickener, in gph?

ANS_____

3. The waste activated sludge flow to a basket centrifuge is 65 gpm. The basket run time is 20 minutes until the basket is full of solids. If it takes 1.5 minutes to skim the solids out of the unit, what is the hydraulic loading rate on the unit in gal/hr?

ANS_____

4. The sludge flow to a basket centrifuge is 76,000 gpd. The basket run time is 23 minutes until the flow to the unit must be stopped for the skimming operation. If skimming takes 2 minutes, what is the hydraulic loading on the unit in gal/hr?

ANS_____

5. A scroll centrifuge receives a waste activated sludge flow of 105,000 gpd with a solids concentration of 7500 mg/L. What is the solids loading in lbs/hr to the centrifuge?

ANS_____

6. The sludge flow to a basket thickener is 70 gpm with a solids concentration of 7400 mg/L. The basket operates 25 minutes before the flow must be stopped to the unit during the 2-minute skimming operation. What is the solids loading in lbs/hr to the centrifuge?

ANS_____

7. A basket centrifuge with a 30-cu ft capacity receives a flow of 70 gpm. The influent sludge solids concentration is 7300 mg/L. The average solids concentration within the basket is 6.8%. What is the feed time (minutes) for the centrifuge?

ANS_____

8. A basket centrifuge thickener has a capacity of 20 cu ft. The 50-gpm sludge flow to the thickener has a solids concentration of 7400 mg/L. The average solids concentration within the basket is 8%. What is the feed time (minutes) for the centrifuge?

ANS_____

PRACTICE PROBLEMS 15.6: Centrifuge Thickening—Cont'd

9. The influent sludge solids concentration to a disc centrifuge is 7800 mg/L. If the sludge solids concentration of the centrifuge effluent (centrate) is 700 mg/L, what is the sludge solids removal efficiency?

ANS_____

10. The influent sludge to a scroll centrifuge has a sludge solids concentration of 9000 mg/L. If the centrifuge effluent has a sludge solids concentration of 0.25%, what is the sludge solids removal efficiency?

ANS_____

11. A total of 15 cu ft of skimmed sludge and 3.5 cu ft of knifed sludge is removed from a basket centrifuge. If the skimmed sludge has a solids concentration of 4.2% and the knifed sludge has a solids concentration of 7.5%, what is the percent solids concentration of the sludge mixture?

ANS_____

12. A total of 11 cu ft of skimmed sludge and 3 cu ft of knifed sludge is removed from a basket centrifuge. If the skimmed sludge has a solids concentration of 3.5% and the knifed sludge has a solids concentration of 7.8%, what is the percent solids concentration of the sludge mixture?

ANS_____

Chapter 15—Achievement Test

1. The total weight of a sludge sample is 30 grams (sample weight only, not including the weight of the dish). If the weight of the solids after drying is 0.58 grams, what is the percent total solids of the sludge?

ANS_____

2. The suspended solids content of the primary influent is 295 mg/L and that of the primary effluent is 119 mg/L. How many pounds of solids are produced during a day that the flow is 3.2 MGD?

ANS_____

3. A primary clarifier sludge has a 3.8% solids content. If 3700 gpd primary sludge is pumped to a thickener and the thickened sludge has a solids content of 7%, what would be the expected gpd flow of thickened sludge? (Assume both sludges weigh 8.34 lbs/gal.)

ANS_____

4. A total of 9400 gallons of sludge is pumped to a digester daily. If the sludge has a 4.7% solids content, how many lbs/day solids are pumped to the digester? (Assume the sludge weighs 8.34 lbs/gal.)

ANS_____

5. The 2.92-MGD influent to a secondary system has a BOD concentration of 160 mg/L. The secondary effluent contains 35 mg/L BOD. If the bacteria growth rate, Y-value, for this plant is 0.5 lbs SS/lb BOD removed, what is the estimated pounds of dry solids are produced each day by the secondary system?

ANS_____

6. A gravity thickener 40 ft in diameter has a sludge blanket depth of 4 ft. If sludge is pumped from the bottom of the thickener at the rate of 30 gpm, what is the sludge detention time (in hours) in the thickener? (Round to the nearest tenth.)

ANS_____

7. The solids concentration of the influent flow to a gravity thickener is 2.9%. If the sludge withdrawn from the bottom of the thickener has a solids concentration of 7.5%, what is the concentration factor?

ANS_____

8. The sludge flow to a 35-ft diameter dissolved air thickener is 140,000 gpd. If the influent waste activated sludge has a suspended solids concentration of 7850 mg/L, what is the solids loading rate in lbs/hr/sq ft? (Round to the nearest tenth.)

ANS_____

9. A 70-ft diameter DAF thickener receives a sludge flow with a solids concentration of 6900 mg/L. If the effluent solids concentration is 220 mg/L, what is the solids removal efficiency?

ANS_____

Chapter 15—Achievement Test—Cont'd

10. The primary sludge flow to a gravity thickener is 65 gpm. This is blended with a 95 gpm secondary effluent flow. If the thickener has a diameter of 25 ft, what is the hydraulic loading rate in gpd/sq ft?

ANS_____

11. What is the efficiency of a gravity thickener if the influent flow to the thickener has a sludge solids concentration of 3.1% and the effluent flow has a sludge solids concentration of 0.3%.

ANS_____

12. The air supplied to a dissolved air flotation thickener is 8 cfm. What is this air supply expressed as lbs/hr?

ANS_____

13. Given the data below, (a) determine whether the sludge blanket in the gravity thickener will increase, decrease, or remain the same; and (b) if there is an increase or decrease, how many lbs/day change will this be?

Sludge Pumped to Thickener—105 gpm
Thickened Sludge Pumped from Thickener—45 gpm
Primary Sludge Solids—3%
Thickened Sludge Solids—7.4%
Thickener Effluent Suspended Solids—680 mg/L

ANS_____

14. What is the solids loading on a gravity thickener, in lbs/day/sq ft, if the primary sludge flow to the 30-ft diameter gravity thickener is 50 gpm with a solids concentration of 3.8%? (Round to the nearest tenth.)

ANS_____

15. A dissolved air flotation unit is 50 ft long and 15 ft wide. If the unit receives a sludge flow of 180,000 gpd, what is the hydraulic loading rate in gpm/sq ft? (Round to the nearest tenth.)

ANS_____

16. If solids are being stored at a rate of 9200 lbs/day in a 25-ft diameter gravity thickener, how many hours will it take the sludge blanket to rise 2.4 ft? The solids concentration of the thickened sludge is 6.7%. (Round to the nearest tenth.)

ANS_____

17. The waste activated sludge flow to a scroll centrifuge thickener is 80,000 gpd. What is the hydraulic loading rate on the thickener in gph?

ANS_____

Chapter 15—Achievement Test—Cont'd

18. The sludge flow to a DAF thickener is 100 gpm. The solids concentration of the sludge is 0.72%. If the air supplied to the DAF unit is 5 cfm, what is the air-to-solids ratio? (Round to the nearest hundredth.)

ANS_____

19. The desired percent recycle for a DAF unit is 110%. If the sludge flow to the thickener is 70 gpm, what should the recycle flow be in MGD?

ANS_____

20. The sludge flow to a basket centrifuge is 77,000 gpd. The basket run time is 28 minutes until the flow to the unit must be stopped for the skimming operation. If skimming takes 2 minutes, what is the hydraulic loading on the unit in gal/hr?

ANS_____

21. After several hours of startup of a gravity thickener, the sludge blanket level is measured at 2.3 ft. The desired sludge blanket level is 5 ft. If the sludge solids are entering the thickener at a rate of 45 lbs/min, what is the desired sludge withdrawal rate in gpm? The thickened sludge solids concentration is 7%.

ANS_____

22. A scroll centrifuge receives a waste activated sludge flow of 108,000 gpd with a solids concentration of 6950 mg/L. What is the solids loading to the centrifuge in lbs/hr?

ANS_____

23. A basket centrifuge with a 30 cu ft capacity receives a flow of 60 gpm. The influent sludge solids concentration is 7100 mg/L. The average solids concentration within the basket is 6.4%. What is the feed time for the centrifuge, in minutes?

ANS_____

24. The sludge flow to a basket thickener is 90 gpm with a solids concentration of 7700 mg/L. The basket operates 23 minutes before the flow must be stopped to the unit during the 1.5-minute skimming operation. What is the solids loading to the centrifuge in lbs/hr?

ANS_____

25. A total of 14 cu ft of skimmed sludge and 4 cu ft of knifed sludge is removed from a basket centrifuge. If the skimmed sludge has a solids concentration of 3.6% and the knifed sludge has a solids concentration of 7.6%, what is the solids concentration of the sludge mixture?

ANS_____

16 *Sludge Digestion*

PRACTICE PROBLEMS 16.1: Mixing Different Percent Solids Sludges

1. A primary sludge flow of 4120 gpd with a solids content of 5.6% is mixed with a thickened secondary sludge flow of 6740 gpd with a solids content of 3.4%. What is the percent solids content of the mixed sludge flow? Assume both sludges weigh 8.34 lbs/gal. (Round to the nearest tenth.)

ANS_____

2. Primary and thickened secondary sludges are to be mixed and sent to the digester. The 3430-gpd primary sludge has a solids content of 5.4% and the 5190-gpd thickened secondary sludge has a solids content of 3.9%. What would be the percent solids content of the mixed sludge? Assume both sludges weigh 8.34 lbs/gal. (Round to the nearest tenth.)

ANS_____

3. A primary sludge flow of 3840 gpd with a solids content of 6.1% is mixed with a thickened secondary sludge flow of 6670 gpd with a solids content of 4.6%. What is the percent solids of the combined sludge flow? Assume both sludges weigh 8.34 lbs/gal. (Round to the nearest tenth.)

ANS_____

4. Primary and thickened secondary sludges are to be mixed and sent to the digester. The 2470 gpd primary sludge has a solids content of 4.2%, and the 3510 gpd thickened secondary sludge has a solids content of 5.8%. What would be the percent solids content of the mixed sludge? Assume the 4.2% sludge weighs 8.34 lbs/gal and the 5.8% sludge weighs 8.62 lbs/gal. (Round to the nearest tenth.)

ANS_____

PRACTICE PROBLEMS 16.2: Sludge Volume Pumped

1. A piston pump discharges a total of 0.8 gallons per stroke (or revolution). If the pump operates at 30 strokes per minute, what is the gpm pumping rate? (Assume the piston is 100% efficient and displaces 100% of its volume each stroke)

ANS_____

2. A sludge pump has a bore of 10 inches and a stroke length of 3 inches. If the pump operates at 32 strokes per minute, how many gpm are pumped? (Assume the piston is 100% efficient and displaces 100% of its volume each stroke.)

ANS_____

3. A sludge pump has a bore of 8 inches and a stroke setting of 4 inches. The pump operates at 45 strokes per minute. If the pump operates a total of 120 minutes during a 24-hour period, what is the gpd pumping rate? (Assume the piston is 100% efficient.)

ANS_____

4. A sludge pump has a bore of 12 inches and a stroke setting of 5 inches. The pump operates at 35 strokes per minute. If the pump operates a total of 150 minutes during a 24-hour period, what is the gpd pumping rate? (Assume the piston is 100% efficient.)

ANS_____

PRACTICE PROBLEMS 16.3: Sludge Pump Operating Time

1. The flow to a primary clarifier is 2.3 MGD. The influent suspended solids concentration is 220 mg/L and the effluent suspended solids concentration is 108 mg/L. If the sludge to be removed from the clarifier has a solids content of 3.3% and the sludge pumping rate is 35 gpm, how many minutes per hour should the pump operate?

ANS_____

2. The suspended solids concentration of the 1,950,000-gpd influent flow to a primary clarifier is 205 mg/L. The primary clarifier effluent flow suspended solids concentration is 95 mg/L. If the sludge to be removed from the clarifier has a solids content of 3.8% and the sludge pumping rate is 30 gpm, how many minutes per hour should the pump operate?

ANS_____

3. A primary clarifier receives a flow of 3,450,000 gpd with a suspended solids concentration of 222 mg/L. The clarifier effluent has a suspended solids concentration of 98 mg/L. If the sludge to be removed from the clarifier has a solids content of 4.1%, and the sludge pumping rate is 40 gpm, how many minutes per hour should the pump operate?

ANS_____

4. The flow to a primary clarifier is 1.3 MGD with a suspended solids concentration of 217 mg/L. The clarifier effluent suspended solids concentration is 94 mg/L. The sludge to be removed from the clarifier has a solids content of 3.5%. If the sludge pumping rate is 35 gpm, how many minutes per hour should the pump operate?

ANS_____

PRACTICE PROBLEMS 16.4: Volatile Solids to the Digester

1. If 8,450 lbs/day of solids with a volatile solids content of 67% are sent to the digester, how many lbs/day volatile solids are sent to the digester?

ANS_____

2. If 2,780 lbs/day of solids with a volatile solids content of 65% are sent to the digester, how many lbs/day volatile solids are sent to the digester?

ANS_____

3. A total of 3,630 gpd of sludge is to be pumped to the digester. If the sludge has a 5.7% solids content with 71% volatile solids, how many lbs/day volatile solids are pumped to the digester? (Assume the sludge weighs 8.34 lbs/gal.)

ANS_____

4. The sludge has a 6% solids content with 69% volatile solids. If a total of 5,015 gpd of sludge is to be pumped to the digester, how many lbs/day volatile solids are pumped to the digester?

ANS_____

PRACTICE PROBLEMS 16.5: Seed Sludge Based On Digester Capacity

1. A digester has a capacity of 293,590 gallons. If the digester seed sludge is to be 25% of the digester capacity, how many gallons of seed sludge will be required?

ANS_____

2. A 40-ft diameter digester has a side water depth of 23 ft. If the seed sludge to be used is 22% of the tank capacity, how many gallons of seed sludge will be required?

ANS_____

3. A 40-ft diameter digester has a side water depth of 21 ft. If 61,590 gallons of seed sludge are to be used in starting up the digester, what percent of the digester volume will be seed sludge?

ANS_____

4. A digester, 50 ft in diameter, has a side water depth of 19 ft. If the digester seed sludge is to be 20% of the digester capacity, how many gallons of seed sludge will be required?

ANS_____

PRACTICE PROBLEMS 16.6: Seed Sludge Based on Volatile Solids Loading

1. A total of 65,570 lbs/day of sludge is pumped to a 100,000-gal digester. The sludge being pumped to the digester has total solids content of 5.2% and a volatile solids content of 69%. The sludge in the digester has a solids content of 6.2% with a 55% volatile solids content. What is the volatile solids loading on the digester in lbs VS added/day/lb VS in digester? (Round to the nearest hundredth.)

ANS_____

2. A total of 21,190 gal of digested sludge are in a digester. The digested sludge contains 6.1% total solids and 56% volatile solids. To maintain a VS loading ratio of 0.05 lbs VS added/day/lb VS under digestion, how many lbs VS may enter the digester daily?

ANS_____

3. A total of 59,880 lbs/day sludge is pumped to a 95,000-gal digester. The sludge pumped to the digester has a total solids content of 5.2% and a volatile solids content of 68%. The sludge in the digester has a solids content of 6% with a 59% volatile solids content. What is the volatile solids loading on the digester in lbs VS added/day/lb VS in digester?

ANS_____

4. The raw sludge flow to the new digester is expected to be 910 gpd. The raw sludge contains 5.6% solids and 70% volatile solids. The desired VS loading ratio is 0.09 lbs VS added/day/lb VS in the digester. How many gallons of seed sludge will be required if the seed sludge contains 8.5% solids with a 54% volatile solids content? (Assume the seed sludge weighs 8.85 lbs/gal.)

ANS_____

PRACTICE PROBLEMS 16.7: Digester Loading Rate, lbs VS/day/cu ft

1. A digester 40 ft in diameter with a water depth of 21 ft receives 85,460 lbs/day raw sludge. If the sludge contains 6% solids with 71% volatile matter, what is the digester loading in lbs VS added/day/cu ft volume? (Round to the nearest hundredth.)

    ANS_____

2. A digester 50 ft in diameter with a liquid level of 22 ft receives 38,120 gpd of sludge with 5.7% solids and 69% volatile solids. What is the digester loading in lbs VS added/day/1000 cu ft? (Assume the sludge weighs 8.34 lbs/gal.)

    ANS_____

3. What is the digester loading in lbs VS added/day/1000 cu ft if a digester, 45 ft in diameter with a liquid level of 20 ft, receives 29,900 gpd of sludge with 5.8% solids and 70% volatile solids? (Assume the sludge weighs 8.34 lbs/gal.)

    ANS_____

4. A digester, 40 ft in diameter with a liquid level of 19 ft, receives 17,000 gpd sludge with 5.2% solids and 70% volatile solids. What is the digester loading in lbs VS added/day/1000 cu ft?

ANS_____

PRACTICE PROBLEMS 16.8: Digester Sludge to Remain in Storage

1. A total of 2800 gpd sludge is pumped to a digester. If the sludge has a total solids content of 5.6% and a volatile solids concentration of 68%, how many pounds of digested sludge should be in the digester for this load? Assume the sludge weighs 8.34 lbs/gal. (Use a ratio of 1 lb VS added/day per 10 lbs of digested sludge.)

ANS_____

2. A total of 6200 gpd of sludge is pumped to a digester. The sludge has a solids concentration of 6% and a volatile solids content of 72%. How many pound of digested sludge should be in the digester for this load? Assume the sludge weighs 8.34 lbs/gal. (Use a ratio of 1 lb VS added/day per 10 lbs of digested sludge.)

ANS_____

3. The sludge pumped to a digester has a solids concentration of 6.3% and a volatile solids content of 69%. If a total of 5100 gpd of sludge is pumped to the digester, how many pound of digested sludge should be in the digester for this load? Assume the sludge weighs 8.34 lbs/gal. (Use a ratio of 1 lb VS added/day per 10 lbs of digested sludge.)

ANS_____

4. A digester receives a flow of 3900 gallons of sludge during a 24-hour period. If the sludge has a solids content of 7% and a volatile solids concentration of 70%, how many pounds of digested sludge should be in the digester for this load? Assume the sludge weighs 8.34 lbs/gal. (Use a ratio of 1 lb VS added/day per 10 lbs of digested sludge.)

ANS_____

PRACTICE PROBLEMS 16.9: Volatile Acids/Alkalinity Ratio

1. The volatile acids concentration of the sludge in the anaerobic digester is 172 mg/L. If the measured alkalinity is 2130 mg/L, what is the VA/Alkalinity ratio?

ANS_____

2. The volatile acid concentration of the sludge in the anaerobic digester is 150 mg/L. If the measured alkalinity is 2490 mg/L, what is the VA/Alkalinity ratio?

ANS_____

3. The measured alkalinity is 2320 mg/L. If the volatile acids concentration of the sludge in the anaerobic digester is 146 mg/L, what is the VA/Alkalinity ratio?

ANS_____

4. The measured alkalinity is 2560 mg/L. If the volatile acid concentration of the sludge in the anaerobic digester is 181 mg/L, what is the VA/Alkalinity ratio?

ANS_____

PRACTICE PROBLEMS 16.10: Lime Required For Neutralization

1. To neutralize a sour digester, one mg/L of lime is to be added for every mg/L of volatile acids in the digester sludge. If the digester contains 242,000 gal of sludge with a volatile acid (VA) level of 2270 mg/L, how many pounds of lime should be added?

ANS_____

2. To neutralize a sour digester, one mg/L of lime is to be added for every mg/L of volatile acids in the digester sludge. If the digester contains 195,000 gal of sludge with a volatile acid (VA) level of 1990 mg/L, how many pounds of lime should be added?

ANS_____

3. The digester contains 225,000 gal of sludge with a volatile acid (VA) level of 2510 mg/L. To neutralize a sour digester, one mg/L of lime is to be added for every mg/L of volatile acids in the digester sludge. How many pounds of lime should be added?

ANS_____

4. The digester sludge is found to have a volatile acids content of 2360 mg/L. If the digester volume is 180,000 gallons, how many pounds of lime will be required for neutralization?

ANS_____

PRACTICE PROBLEMS 16.11: Percent Volatile Solids Reduction

1. The sludge entering a digester has a volatile solids content of 66%. The sludge leaving the digester has a volatile solids content of 51%. What is the percent volatile solids reduction?

ANS_____

2. The sludge leaving the digester has a volatile solids content of 53%. The sludge entering a digester has a volatile solids content of 72%. What is the percent volatile solids reduction?

ANS_____

3. The raw sludge to a digester has a volatile solids content of 69%. The digested sludge volatile solids content is 53%. What is the percent volatile solids reduction?

ANS_____

4. The digested sludge volatile solids content is 52%. The raw sludge to a digester has a volatile solids content of 67%. What is the percent volatile solids reduction?

ANS_____

PRACTICE PROBLEMS 16.12:   Volatile Solids Destroyed, lbs VS/cu ft

1. A flow of 3700 gpd sludge is pumped to a 35,000 cu ft digester. The solids concentration of the sludge is 6.1% with a volatile solids content of 71%. If the volatile solids reduction during digestion is 55%, how many lbs/day volatile soids are destroyed per cu ft of digester capacity? Assume the sludge weighs 8.34 lbs/gal. (Round to the nearest hundredth.)

ANS_____

2. A flow of 4450 gpd sludge is pumped to a 33,000 cu ft digester. The solids concentration of the sludge is 6% with a volatile solids content of 68%. If the volatile solids reduction during digestion is 56%, how many lbs/day volatile soids are destroyed per cu ft of digester capacity? Assume the sludge weighs 8.34 lbs/gal. (Round to the nearest hundredth.)

ANS_____

3. A 50-ft diameter digester receives a sludge flow of 2800 gpd with a solids content of 5.8% and a volatile solids concentration of 70%. The volatile solids reduction during digestion is 54%. The digester operates at a level of 20 ft. What is the lbs/day volatile solids reduction per cu ft of digester capacity? Assume the sludge weighs 8.34 lbs/gal. (Round to the nearest hundredth.)

ANS_____

4. The sludge flow to a 45-ft diameter digester is 3000 gpd with a solids concentration of 6.3% and a volatile solids concentration of 67%. The digester is operated at a depth of 18 ft. If the volatile solids reduction during digestion is 58%, what is the lbs/day volatile solids reduction per 1000 cu ft of digester capacity? Assume the sludge weighs 8.34 lbs/gal.

ANS_____

PRACTICE PROBLEMS 16.13: Digester Gas Production

1. A digester gas meter reading indicates an average of 6620 cu ft of gas is produced per day. If a total of 520 lbs/day volatile solids are destroyed, what is the digester gas production in cu ft gas/lb VS destroyed? (Round to the nearest tenth.)

ANS_____

2. A total of 2060 lbs of volatile solids are pumped to the digester daily. If the percent reduction of volatile solids due to digestion is 57% and the average gas production for the day is 19,150 cu ft, what is the daily gas production in cu ft/lb VS destroyed? (Round to the nearest tenth.)

ANS_____

3. The total of 574 lbs/day volatile solids are destroyed. If a digester gas meter reading indicates an average of 8610 cu ft of gas is produced per day, what is the digester gas production in cu ft gas/lb VS destroyed?

ANS_____

4. The percent reduction of volatile solids due to digestion is 56% and the average gas production for the day is 25,640 cu ft. If a total of 3270 lbs of volatile solids are pumped to the digester daily, what is the daily gas production in cu ft/lb VS destroyed? (Round to the nearest tenth.)

ANS_____

PRACTICE PROBLEMS 16.14: Solids Balance

1. Given the following data, calculate the solids balance for the digester.

<u>Sludge to Digester</u>

Raw Sludge—27,100 lbs/day
% Solids—5.8%
% Volatile Solids—68%

<u>Sludge After Digestion</u>

Digested Sludge
% Solids—4.5%
% Volatile Solids—56%

2. Given the following data, calculate the solids balance for the digester.

<u>Sludge to Digester</u>

Raw Sludge—28,000 lbs/day
% Solids—6%
% Volatile Solids—71%

<u>Sludge After Digestion</u>

Digested Sludge
% Solids—4.3%
% Volatile Solids—58%

PRACTICE PROBLEMS 16.15: Digestion Time

1. A 50-ft diameter aerobic digester has a side water depth of 10 ft. The sludge flow to the digester is 8900 gpd. Calculate the hydraulic digestion time, in days. (Round to the nearest tenth.)

ANS_____

2. An aerobic digester 40 ft in diameter has a side water depth of 12 ft. The sludge flow to the digester is 8100 gpd. Calculate the hydraulic digestion time in days. (Round to the nearest tenth.)

ANS_____

3. An aerobic digester is 90 ft long, 20 ft wide and has a side water depth of 10 ft. If the sludge flow to the digester is 7600 gpd, what is the hydraulic digestion time in days? (Round to the nearest tenth.)

ANS_____

4. A sludge flow of 10,000 gpd has a solids content of 3.2%. As a result of thickening, the sludge flow is reduced to 5300 gpd with a 6% solids content. Compare the digestion times for the two different sludge flows to a digester 35-ft in diameter with a side water depth of 10 ft.

ANS_____

PRACTICE PROBLEMS 16.16: Air Requirements and Oxygen Uptake

1. The desired air supply rate for an aerobic digester was determined to be 0.04 cfm/cu ft digester capacity. What is the total cfm air required if the digester is 85 ft long, 25 ft wide, with a side water depth of 10 ft?

ANS_____

2. An aerobic digester is 60 ft in diameter with a side water depth of 10 ft. If the desired air supply for this digester was determined to be 45 cfm/1000-cu ft digester capacity, what is the total cfm air required for this digester?

ANS_____

3. The dissolved air concentrations recorded during a 5-minute test of an air-saturated sample of aerobic digester sludge are given below. Calculate the oxygen uptake, in mg/L/hr.

Elapsed Time, min	D.O., mg/L	Elapsed Time, min	D.O., mg/L
At Start	6.7	3 min	4.4
1 min	5.7	4 min	3.7
2 min	5.2	5 min	3.1

ANS_____

4. The dissolved air concentrations recorded during a 5-minute test of an air-saturated sample of aerobic digester sludge are given below. Calculate the oxygen uptake, in mg/L/hr.

Elapsed Time, min	D.O., mg/L	Elapsed Time, min	D.O., mg/L
At Start	7.5	3 min	4.6
1 min	6.6	4 min	4.1
2 min	5.3	5 min	3.5

ANS_____

PRACTICE PROBLEMS 16.17: pH Adjustments Using Jar Tests

1. Jar tests indicate that 25 mg of caustic are required to raise the pH of the one-liter sludge sample to 7.0. If the digester volume is 110,000 gallons, how many pounds of caustic will be required for pH adjustment? (Round to the nearest tenth.)

ANS_____

2. Jar tests indicate that 17 mg of caustic are required to raise the pH of the one-liter sludge sample to 7.0. If the digester volume is 147,000 gallons, how many pounds of caustic will be required for pH adjustment? (Round to the nearest tenth.)

ANS_____

3. A two-liter sample of digester sludge is used to determine the required caustic dosage for pH adjustment. If 62 mg of caustic are required for pH adjustment in the jar test, and the digester volume is 53,000 gallons, how many pounds of caustic will be required for pH adjustment? (Round to the nearest tenth.)

ANS_____

4. A 2-liter sample of digested sludge is used to determine the required caustic dosage for pH adjustment. A total of 80 mg caustic were used in the jar test. The aerobic digester is 50 ft in diameter with a side water depth of 12 ft. How many pounds of caustic are required for pH adjustment of the digester?

ANS_____

Chapter 16—Achievement Test

1. A total of 8000 gpd sludge is pumped to a digester. If the sludge has a solids content of 5.8% and a volatile solids concentration of 65%, how many pound of digested sludge should be in the digester for this load? (Use a ratio of 1 lb VS/day per 10 lbs of digested sludge.)

ANS_____

2. If 4200 lbs/day solids with a volatile solids content of 69% are sent to the digester, how many pounds of volatile solids are sent to the digester daily?

ANS_____

3. What is the digester loading in lbs VS added/day/1000 cu ft if a digester, 50 ft in diameter, with a liquid level of 21 ft receives 12,620 gpd of sludge with 5.1% solids and 65% volatile solids?

ANS_____

4. A primary sludge flow of 3930 gpd with a solids content of 5.2% is mixed with a thickened secondary sludge flow of 5660 gpd with a solids content of 3.1%. What is the percent solids content of the mixed sludge flow? (Assume both sludges weigh 8.34 lbs/gal.)

ANS_____

5. A sludge pump has a bore of 8 inches and a stroke length of 5 inches. The counter indicates a total of 3400 revolutions during a 24-hour period. What is the pumping rate in gpd? (Assume the piston is 100% efficient.)

ANS_____

6. A 50-ft diameter digester has a typical side water depth of 22 ft. If 87,200 gallons seed sludge are to be used in starting up the digester, what percent of the digester volume will be seed sludge?

ANS_____

7. A flow of 3600 gpd sludge is pumped to a 35,000-cu ft digester. The solids content of the sludge is 3.9% with a volatile solids content of 71%. If the volatile solids reduction during digestion is 56%, how many lbs/day volatile solids are destroyed per cu ft of digester capacity? Assume the sludge weighs 8.34 lbs/gal.

ANS_____

8. The volatile acid concentration of the sludge in the anaerobic digester is 153 mg/L. If the measured alkalinity is 2260 mg/L, what is the VA/Alkalinity ratio? (Round to the nearest hundredth.)

ANS_____

Chapter 16—Achievement Test—Cont'd

9. To neutralize a sour digester, one mg/L of lime is to be added for every mg/L of volatile acid in the digester sludge. If the digester contains 230,000 gal of sludge with a volatile acid (VA) level of 2130 mg/L, how many pounds of lime should be added?

ANS_____

10. A 45-ft diameter digester has a typical water depth of 20 ft. If the seed sludge to be used is 22% of the tank capacity, how many gallons of seed sludge will be required?

ANS_____

11. A total of 4260 gpd of sludge is to be pumped to the digester. If the sludge has a 5.3% solids content with 70% volatile solids, how many lbs/day volatile solids are pumped to the digester? (Assume the sludge weighs 8.34 lbs/gal.)

ANS_____

12. Primary and thickened secondary sludges are to be mixed and sent to the digester. The 2820-gpd primary sludge has a solids content of 5.7%, and the 4650-gpd thickened secondary sludge has a solids content of 3.6%. What would be the percent solids content of the mixed sludge? (Assume both sludges weigh 8.34 lbs/gal.)

ANS_____

13. The measured alkalinity is 2490 mg/*L*. If the volatile acids concentration of the sludge in the anaerobic digester is 160 mg/*L*, what is the VA/Alkalinity ratio?

ANS_____

14. A total of 41,820 lbs/day sludge is pumped to a 90,000-gal digester. The sludge being pumped to the digester has total solids content of 5% and volatile solids content of 62%. The sludge in the digester has a solids content of 6.1% with a 56% volatile solids content. What is the volatile solids loading on the digester in lbs VS added /day/lb VS in digester?

ANS_____

15. A sludge pump has a bore of 10 inches and a stroke length of 4 inches. If the pump operates at 30 strokes (or revolutions) per minute, how many gpm are pumped? (Assume the piston is 100% efficient and displaces 100% of its volume each stroke.)

ANS_____

16. What is the digester loading in lbs VS added/day/1000 cu ft if a digester, 45 ft in diameter with a liquid level of 23 ft, receives 19,790 gpd of sludge with 6% solids and 68% volatile solids?

ANS_____

Chapter 16—Achievement Test—Cont'd

17. The digester sludge is found to have a volatile acids content of 2100 mg/L. If the digester volume is 0.2 MG, how many pounds of lime will be required for neutralization?

ANS_____

18. The sludge entering a digester has a volatile solids content of 69%. The sludge leaving the digester has a volatile solids content of 53%. What is the percent volatile solids reduction?

ANS_____

19. A digester gas meter reading indicates an average of 6870 cu ft of gas is produced per day. If a total of 560 lb/day volatile solids are destroyed, what is the digester gas production in cu ft gas/lb VS destroyed? (Round to the nearest tenth.)

ANS_____

20. The raw sludge flow to a new digester is expected to be 1170 gpd. The raw sludge contains 4.8% solids and 68% volatile solids. The desired VS loading ratio is 0.08 lbs VS added/lb VS in the digester. How many gallons of seed sludge will be required if the seed sludge contains 7.2% solids with a 54% volatile solids content? (Assume the raw sludge weighs 8.34 lbs/gal and the seed sludge weighs 8.5 lbs/gal.)

ANS_____

21. The sludge leaving the digester has a volatile solids content of 55%. The sludge entering a digester has a volatile solids content of 71%. What is the percent volatile solids reduction?

ANS_____

22. A 50-ft diameter aerobic digester has a side water depth of 10 ft. The sludge flow to the digester is 9500 gpd. Calculate the digestion time (hydraulic) in days.

ANS_____

Chapter 16—Achievement Test—Cont'd

23. A total of 2580 lbs of volatile solids are pumped to the digester daily. If the percent reduction of volatile solids due to digestion is 58% and the average gas production for the day is 21,300 cu ft, what is the daily gas production in cu ft/lb VS destroyed?

ANS_____

24. The sludge flow to a 45-ft diameter digester is 3100 gpd with a solids content of 6.3% and a volatile solids concentration of 66%. The digester is operated at a depth of 20 ft. If the volatile solids reduction during digestion is 56%, what is the lbs/day volatile solids reduction per 1000 cu ft of digester capacity?

ANS_____

25. Given the following data calculate the solids balance for the digester.

Sludge to Digester

Raw Sludge—28,300 lbs/day
% Solids—5.9%
% Volatile Solids—69%

Sludge After Digestion

Digested Sludge
% Solids—4%
% Volatile Solids—52%

Chapter 16—Achievement Test—Cont'd

26. The desired air supply rate for an aerobic digester is determined to be 0.04 cfm/cu ft digester capacity. What is the total cfm air required if the digester is 85 ft long, 25 ft wide and has a side water depth of 10 ft?

ANS_____

27. Jar testing indicates that 24 mg of caustic are required to raise the pH of the one-liter sample to 7.0. If the digester volume is 140,000 gallons, how many pounds of caustic will be required for pH adjustment?

ANS_____

28. Dissolved air concentrations are taken on an air-saturated sample of digested sludge at one-minute intervals. Given the results below, calculate the oxygen uptake, in mg/L/hr.

Elapsed Time, min	D.O., mg/L	Elapsed Time, min	D.O., mg/L
At Start	7.9	3 min	5.3
1 min	6.8	4 min	4.6
2 min	6.1	5 min	3.9

ANS_____

29. The flow to a primary clarifier is 2.4 MGD. The influent suspended solids concentration is 225 mg/*L* and the effluent suspended solids concentration is 103 mg/*L*. If the sludge to be removed from the clarifier has a solids content of 3.1% and the sludge pumping rate is 25 gpm, how many minutes per hour should the pump operate?

ANS_____

30. A sludge flow of 10,000 gpd has a solids concentration of 2.4%. The solids concentration is increased to 4.5% as a result of thickening and the reduced flow rate is 5300 gpd. Compare the digestion time (hydraulic) for these two different sludge flows. The digester is 30 ft in diameter with a 22 ft operating depth.

ANS_____

17 *Sludge Dewatering and and Disposal*

PRACTICE PROBLEMS 17.1: Filter Press Dewatering

1. A filter press used to dewater digested primary sludge receives a flow of 1000 gallons during a 3-hr period. The sludge has a solids content of 3.6%. If the plate surface area is 130 sq ft, what is the solids loading rate in lbs/hr/sq ft? Assume the sludge weighs 8.34 lbs/gal. (Round to the nearest tenth.)

    ANS_____

2. A filter press used to dewater digested primary sludge receives a flow of 815 gallons during a 2-hr period. The solids content of the sludge is 4%. If the plate surface area is 150 sq ft, what is the solids loading rate in lbs/hr/sq ft? Assume the sludge weighs 8.34 lbs/gal. (Round to the nearest tenth.)

    ANS_____

3. A plate and frame filter press receives a solids loading of 0.75 lbs/hr/sq ft. If the filtration time is 2 hours and the time required to remove the sludge cake and begin sludge feed to the press is 20 minutes, what is the net filter yield in lbs/hr/sq ft? (Round to the nearest tenth.)

    ANS_____

4. A plate and frame filter press receives a flow of 670 gallons of sludge during a 2-hour period. The solids concentration of the sludge is 3.8%. The surface area of the plate is 120 sq ft. If the down time for sludge cake discharge is 25 minutes, what is the net filter yield in lbs/hr/sq ft? Assume the sludge weighs 8.34 lbs/gal. (Round to the nearest tenth.)

ANS_____

PRACTICE PROBLEMS 17.2: Belt Filter Press Dewatering

1. A 6-feet wide belt press receives a flow of 130 gpm of primary sludge. What is the hydraulic loading rate in gpm/ft?

ANS_____

2. The amount of sludge to be dewatered by the belt filter press is 20,100 lbs/day. If the belt filter press is to be operated 14 hours each day, what should the lbs/hr sludge feed rate be to the press?

ANS_____

3. The amount of sludge to be dewatered is by a belt filter press is 22,600 lbs/day. If the maximum feed rate that still provides an acceptable cake is 1700 lbs/hr, how many hours per day should the belt remain in operation?

ANS_____

4. The sludge feed to a belt filter press is 150 gpm. If the total suspended solids concentration of the feed is 4.2%, what is the solids loading rate, in lbs/hr? Assume the sludge weighs 8.34 lbs/gal.

ANS_____

5. The flocculant concentration for a belt filter press is 0.8%. If the flocculant feed rate is 3 gpm, what is the flocculant feed rate in lbs/hr? (Assume the flow is steady and continuous and assume the flocculant weighs 8.34 lbs/gal.)

ANS_____

6. Using the results from problems 4 and 5, calculate the flocculant dose in pounds per ton of solids treated. (Round to the nearest tenth.)

ANS_____

7. Given the solids concentrations below, calculate the percent solids recovery of the belt filter press.

Feed Sludge TSS, %—3.7% : X_S
Return Flow TSS, %—0.039% : X_R
Cake TS, %—15% : X_C

ANS_____

ANS_____

PRACTICE PROBLEMS 17.3: Vacuum Filter Dewatering

1. Digested sludge is applied to a vacuum filter at a rate of 85 gpm, with a solids concentration of 4.8%. If the vacuum filter has a surface area of 310 sq ft, what is the filter loading in lbs/hr/sq ft? Assume the sludge weighs 8.34 lbs/gal. (Round to the nearest tenth.)

ANS_____

2. The wet cake flow from a vacuum filter is 6750 lbs/hr. If the filter area is 300 sq ft and the percent solids in the cake is 28%, what is the filter yield in lbs/hr/sq ft? (Round to the nearest tenth.)

ANS_____

3. A total of 5300 lbs/day primary sludge solids are to be processed by a vacuum filter. The vacuum filter yield is 3.2 lbs/hr/sq ft. The solids recovery is 95%. If the area of the filter is 220 sq ft, how many hours per day must the vacuum filter remain in operation to process this much solids? (Round to the nearest tenth.)

ANS_____

4. The total pounds of dry solids pumped to a vacuum filter during a 24-hour period is 17,240 lbs/day. The vacuum filter is operated 10 hrs/day. If the percent solids recovery is 93% and the filter area is 280 sq ft, what is the filter yield in lbs/hr/sq ft?

ANS_____

5. The sludge feed to a vacuum filter is 84,800 lbs/hr, with a solids content of 5.7%. If the wet cake flow is 18,300 lbs/hr, with a 25% solids content, what is the percent solids content, what is the percent solids recovery?

ANS_____

PRACTICE PROBLEMS 17.4: Sand Drying Beds

1. A drying bed is 200 ft long and 20 ft wide. If sludge is applied to a depth of 6 inches, how many gallons of sludge are applied to the drying bed?

ANS_____

2. A drying bed is 230 ft long and 25 ft wide. If sludge is applied to a depth of 8 inches, how many gallons of sludge are applied to the drying bed?

ANS_____

3. A sludge bed is 180 ft long and 25 ft wide. A total of 166,000 lbs of sludge are applied each application of the sand drying bed. The sludge has a solids content of 4.8%. If the drying and removal cycle requires 22 days, what is the solids loading rate in lbs/yr/sq ft?

ANS_____

4. A sludge drying bed is 210 ft long and 25 ft wide. The sludge is applied to a depth of 10 inches. The solids concentration of the sludge is 3.8%. If the drying and removal cycle requires 25 days, what is the solids loading rate to the beds in lbs/yr/sq ft? Assume the sludge weighs 8.34 lbs/gal.

ANS_____

5. Sludge is withdrawn from a digester which has a diameter of 40 feet. If the sludge is drawn down 2.5 feet, how many cu ft will be sent to the drying beds?

ANS_____

6. A 45-feet diameter digester has a drawdown of 16 inches. If the drying bed is 80 ft long and 40 feet wide, (a) how many feet deep will the drying bed be as a result of the drawdown? (Round to the nearest hundredth.) and (b) how many inches is this? (Round to the nearest inch)

ANS_____

ANS_____

PRACTICE PROBLEMS 17.5: Composting

1. If 4500 lbs/day dewatered sludge, with a solids content of 22% solids is mixed with 3700 lbs/day compost, with a 25% moisture content, what is the percent moisture of the blend?

ANS_____

2. If 9,600 lbs/day dewatered sludge, with a solids content of 15% solids is mixed with 10,500 lbs/day compost, with a 28% moisture content, what is the percent moisture of the blend?

ANS_____

3. The total dewatered digested primary sludge produced at a plant is 4600 lbs/day, with a solids content of 18%. The final compost to be used in blending has a moisture content of 28%. How much compost (lbs/day) must be blended with the dewatered sludge to produce a mixture with a moisture content of 45%?

ANS_____

4. Compost is to be blended from wood chips and dewatered sludge. The wood chips are to be mixed with 7.2 cu yds of dewatered sludge at a ratio (by volume) of 3:1. The solids content of the sludge is 18% and the solids content of the wood chips is 53%. If the bulk density of the sludge is 1685 lbs/cu yd and the bulk density of the wood chips is 750 lbs/cu yd, what is the percent solids of the compost blend?

ANS_____

PRACTICE PROBLEMS 17.5: Composting—Cont'd

5. A composting facility has an available capacity of 8000 cu yds. If the composting cycle is 21 days, how many lbs/day wet compost can be processed by this facility? Assume a compost bulk density of 1000 lbs/cu yd.

ANS_____

6. Compost is to be blended from wood chips and dewatered sludge. The wood chips are to be mixed with 10 cu yds of dewatered sludge at a ratio (by volume) of 3:1. The solids content of the sludge is 15% and the solids content of the wood chips is 54%. If the bulk density of the sludge is 1685 lbs/cu yd and the bulk density of the wood chips is 750 lbs/cu yd, what is the percent solids of the compost blend?

ANS_____

7. Given the data listed below, calculate the solids processing capability, lbs/day, of the compost operation.

       Cycle Time—21 days
       Total Available Capacity—7750 cu yds
       % Solids of Wet Sludge—18%
       Mix Ratio (by volume) of Wood Chips to Sludge—3
       Wet Compost Bulk Density—1000 lbs/cu yd
       Wet Sludge Bulk Density—1685 lbs/cu yd
       Wet Wood Chips Bulk Density—750 lbs/cu yd

ANS_____

8. Determine the resulting total solids compost and site capacity (dry solids, tons/wk) for the following conditions: (Use the nomograph provided in Appendix C.)

       Total Solids Content of the Sludge—15%
       Solids Content of Wood Chips—50%   ANS_____
       Mix ratio of Wood Chips to Sludge—3.2
       Cycle Time—28 days

ANS_____

9. Determine the required mix ratio to achieve the desired percent solids compost shown below. Then determine the resulting site capacity (dry solids/wk) for that mix ratio and a 28 day cycle. (Use the nomograph provided in Appendix C.)

       Desired Total Solids of the Compost—41%   ANS_____
       Total Solids Content of the Sludge—17%
       Solids Content of the Wood Chips—60%

ANS_____

Chapter 17—Achievement Test

1. The sludge feed to a belt filter press is 140 gpm. If the total suspended solids concentration of the feed is 4.6%, what is the solids loading rate, in lbs/hr? Assume the sludge weighs 8.34 lbs/gal.

    ANS_____

2. The amount of sludge to be dewatered by the belt filter press is 24,100 lbs/day. If the belt filter press is to be operated 14 hours each day, what should be the sludge feed rate in lbs/hr to the press?

    ANS_____

3. A filter press used to dewater digested primary sludge receives a flow of 750 gallons of sludge during a 2-hr period. The sludge has a solids content of 3.8%. If the plate surface area is 135 sq ft, what is the solids loading rate in lbs/hr/sq ft? Assume the sludge weighs 8.34 lbs/gal. (Round to the nearest tenth.)

    ANS_____

4. The sludge feed rate to a belt filter press is 160 gpm. The total suspended solids concentration of the feed is 4%. The flocculant used for sludge conditioning is a 0.8% concentration, with a feed rate of 2.5 gpm. What is the flocculant dose expressed as pounds flocculant per ton of solids treated?

    ANS_____

5. A plate and frame filter press receives a solids loading of 0.7 lbs/hr/sq ft. If the filtration time is 2 hours and the time required to remove the sludge cake and begin sludge feed to the press is 22 minutes, what is the net filter yield in lbs/hr/sq ft? Assume the sludge weighs 8.34 lbs/gal. (Round to the nearest tenth.)

ANS_____

6. Laboratory tests indicate that the total residue portion of a feed sludge sample is 23,100 mg/L. The total filterable residue is 720 mg/L. On this basis what is the estimated total suspended solids concentration of the sludge sample?

ANS_____

7. Digested sludge is applied to a vacuum filter at a rate of 75 gpm, with a solids concentration of 5.3%. If the vacuum filter has a surface area of 300 sq ft, what is the filter loading in lbs/hr/sq ft? Assume the sludge weighs 8.34 lbs/gal. (Round to the nearest tenth.)

ANS_____

8. The wet cake flow from a vacuum filter is 7400 lbs/hr. If the filter area is 310 sq ft and the percent solids in the cake is 25%, what is the filter yield in lbs/hr/sq ft? (Round to the nearest tenth.)

ANS_____

Chapter 17—Achievement Test—Cont'd

9. Given the solids concentrations below, calculate the percent solids recovery of the belt filter press.

Feed Sludge TSS, %—2.6%   : X_S
Return Flow TSS, %—0.047%   : X_R
Cake TS, %—15%   : X_C

ANS_____

10. The amount of sludge to be dewatered is by a belt filter press is 26,700 lbs/day. If the maximum feed rate which still provides an acceptable cake is 1900 lbs/hr, how many hours per day should the belt remain in operation?

ANS_____

11. A total of 5600 lbs/day primary sludge solids are to be processed by a vacuum filter. The vacuum filter yield is 2.9 lbs/hr/sq ft. The solids recovery is 94%. If the area of the filter is 250 sq ft, how many hours per day must the vacuum filter remain in operation to process this much solids? (Round to the nearest tenth.)

ANS_____

12. A drying bed is 210 ft long and 25 ft wide. If sludge is applied to a depth of 8 inches, how many gallons of sludge are applied to the drying bed?

ANS_____

13. The sludge feed to a vacuum filter is 89,000 lbs/day, with a solids content of 5.1%. If the wet cake flow is 14,100 lbs/day, with a 30% solids content, what is the percent solids recovery?

ANS_____

14. A sludge drying bed is 190 ft long and 20 ft wide. The sludge is applied to a depth of 6 inches. The solids concentration of the sludge is 4.8%. If the drying and removal cycle requires 20 days, what is the solids loading rate to the beds in lbs/yr/sq ft? Assume the sludge weighs 8.34 lbs/gal.

ANS_____

15. A drying bed is 180 ft long and 25 ft wide. If a 45-ft diameter digester has a drawdown of 1 ft, (a) how many feet deep will the drying bed be as a result of the drawdown (rounded to the nearest hundredth), and (b) how many inches is this?

ANS_____

16. A treatment plant produces a total of 6600 lbs/day of dewatered digested primary sludge. The dewatered sludge has a solids concentration of 25%. Final compost to be used in blending has a moisture content of 35%. How much compost (lbs/day) must be blended with the dewatered sludge to produce a mixture with a moisture content of 50%?

ANS_____

Chapter 17—Achievement Test—Cont'd

17. Compost is to be blended from wood chips and dewatered sludge. The wood chips are to be mixed with 6.5 cu yds of dewatered sludge at a ratio of 3:1. The solids content of the sludge is 15% and the solids content of the wood chips is 52%. If the bulk density of the sludge is 1685 lbs/cu yd and the bulk density of the wood chips is 750 lbs/cu yd, what is the percent solids of the compost blend?

ANS_____

18 A composting facility has an available capacity of 6200 cu yds. If the composting cycle is 28 days, how many lbs/day wet compost can be processed by this facility? How many tons/day is this? Assume a compost bulk density of 950 lbs/cu yd.

ANS_____

19. Given the data listed below, calculate the dry sludge processing capability, lbs/day, of the compost operation.

> Cycle time—25 days
> Total available capacity—8000 cu yds
> % Solids of wet sludge—18%
> Mix ratio (by volume) of wood chips to sludge—3.3
> Wet Compost Bulk Density—1000 lbs/cu yd
> Wet Sludge Bulk Density—1685 lbs/cu yd
> Wet Wood Chips Bulk Density—750 lbs/cu yd

ANS_____

20. Determine the resulting total solids compost and site capacity (dry solids, tons/wk) for the following conditions: (Use the nomograph provided in Appendix C.)

> Total solid content of sludge—16%
> Solids content of wood chips—55%
> Mix ratio of wood chips to sludge—3.2
> Cycle time—21 days

ANS_____

18 *Laboratory*

PRACTICE PROBLEMS 18.1: Biochemical Oxygen Demand

1. Given the following information, determine the BOD of the wastewater:

Sample Volume—5 m*L*
BOD Bottle Volume—300 m*L*
Initial DO of Diluted Sample—6 mg/*L*
DO of Diluted Sample—3.5 mg/*L*
(After 5 days)

ANS_____

2. Results from a BOD test are given below. Calculate the BOD of the sample.

Sample Volume—10 m*L*
BOD Bottle Volume—300 m*L*
Initial DO of Diluted Sample—8.3 mg/*L*
DO of Diluted Sample—4.2 mg/*L*
(After 5 days)

ANS_____

3. Given the following primary effluent BOD test results, calculate the 7-day average.

Apr. 10—190 mg/*L*   Apr. 14—210 mg/*L*
Apr. 11—198 mg/*L*   Apr. 15—201 mg/*L*
Apr. 12—205 mg/*L*   Apr. 16—197 mg/*L*
Apr. 13—202 mg/*L*

ANS_____

4. Calculate the 7-day average primary effluent BOD for August 9th, 10th, and 11th, given the BOD test results shown below.

Aug. 1—211 mg/*L*	Aug. 6—220 mg/*L*	Aug. 11—210 mg/*L*
Aug. 2—219 mg/*L*	Aug. 7—225 mg/*L*	Aug. 12—205 mg/*L*
Aug. 3—208 mg/*L*	Aug. 8—206 mg/*L*	Aug. 13—220 mg/*L*
Aug. 4—203 mg/*L*	Aug. 9—195 mg/*L*	Aug. 14—212 mg/*L*
Aug. 5—215 mg/*L*	Aug. 10—198 mg/*L*	Aug. 15—218 mg/*L*

ANS_____

PRACTICE PROBLEMS 18.2: Molarity and Moles

1. If 2.7 moles of solute are dissolved in 0.6 liters solution, what is the molarity of the solution?

ANS_____

2. A 1.5-molar solution is to be prepared. If a total of 800 mL solution is to be prepared, how many moles solute will be required?

ANS_____

3. The atomic weight of calcium is 40. If 26 grams of calcium are used in making up a one-liter solution, how many moles are used? (Round to the nearest tenth.)

ANS_____

4. What is the molarity of a solution that has 0.3 moles solute dissolved in 1600 m*L* solution? (Round to the nearest tenth.)

ANS_____

5. The atomic weights listed for each element of NaOH are given below. How many grams make up a mole of NaOH?

Na: 22.997

O: 16.000

H: 1.008

ANS_____

PRACTICE PROBLEMS 18.3: Normality and Equivalents

1. If 2.1 equivalents of a chemical are dissolved in 1.6 liters solution, what is the normality of the solution? (Round to the nearest tenth.)

ANS_____

2. A 300-m*L* solution contains 1.4 equivalents of a chemical. What is the normality of the solution? (Round to the nearest tenth.)

ANS_____

3. How many milliliters of 0.7N NaOH will react with 750 m*L* of 0.05N HCl?

ANS_____

4. Calcium has an atomic weight of 40. If calcium has two valence electrons, what is the equivalent weight of calcium?

ANS_____

5. The molecular weight of Na_2CO_3 is 106. The net valence is 2. If 105 grams of Na_2CO_3 are dissolved in a solution, how many equivalents are dissolved in the solution?

ANS_____

PRACTICE PROBLEMS 18.4: Settleability

1. The settleability test is conducted on a sample of MLSS. What is the percent settleable solids if 410 milliliters settle in the 2000-m*L* graduate?

ANS_____

2. A 2000-m*L* sample of activated sludge. If the settled sludge is measured as 315 milliliters, what is the percent settleable solids?

ANS_____

3. The settleability test is conducted on a sample of MLSS. What is the percent settleable solids if 390 milliliters settle in the 2000-m*L* graduate?

ANS_____

4. A 2000-mL sample of activated sludge. If the settleable solids are measured as 360 milliliters, what is the percent settled sludge?

ANS_____

PRACTICE PROBLEMS 18.5: Settleable Solids (Imhoff Cone)

1. Calculate the percent removal of settleable solids if the settleable solids of the sedimentation tank influent are 16.5 m*L*/*L* and the settleable solids of the effluent are 0.6 m*L*/*L*.

ANS_____

2. The settleable solids of the raw wastewater is 18 m*L*/*L*. If the settleable solids of the clarifier is 0.9 m*L*/*L*, what is the settleable solids removal efficiency of the clarifier?

ANS_____

3. The settleable solids of the raw wastewater is 20 m*L*/*L*. If the settleable solids of the clarifier is 0.8 m*L*/*L*, what is the settleable solids removal efficiency of the clarifier?

ANS_____

4. A clarifier removes 95% of the settleable solids entering the clarifier. If a total of 19 m*L/L* settleable solids enter the clarifier how, many m*L/L* settleable solids are removed?

ANS_____

PRACTICE PROBLEMS 18.6: Sludge Total Solids and Volatile Solids

1. Given the information below, (a) determine the percent total solids in the sample (rounded to the nearest tenth), and (b) the percent of volatile solids in the sludge sample.

	Sludge (Total Sample)	After Drying	After Burning (Ash)
Weight of Sample & Dish	84.15 g	25.17 g	23.29 g
Weight of Dish (tare wt.)	22.40 g	22.40 g	22.40 g

ANS_____

2. Given the information below, calculate (a) the percent total solids (rounded to the nearest tenth) and (b) the percent volatile solids of the sludge sample.

	Sludge (Total Sample)	After Drying	After Burning (Ash)
Weight of Sample & Dish	75.86 g	22.97 g	21.67 g
Weight of Dish (tare wt.)	21.07 g	21.07 g	21.07 g

ANS_____

3. A 100-m*L* sludge sample has been dried and burned. Given the information below, (a) determine the percent volatile solids content of the sample, and (b) determine the mg/*L* concentration of the volatile solids.

	After Drying	After Burning (Ash)
Weight of Sample & Crucible	22.0173 g	22.0070 g
Weight of Crucible (tare wt.)	22.0024 g	22.0024 g

ANS_____

PRACTICE PROBLEMS 18.7: Suspended Solids and Volatile Suspended Solids (Of Wastewater)

1. Given the following information regarding a primary effluent sample, calculate (a) the mg/*L* suspended solids, and (b) the percent volatile suspended solids of the sample.

	After Drying (Before Burning)	After Burning (Ash)
Weight of Sample & Dish	25.6715 g	25.6701 g
Weight of Dish (Tare Wt.)	25.6670 g	25.6670 g

Sample Volume = 50 m*L*

ANS_____

2. Given the following information regarding a treatment plant influent sample, a) calculate the mg/*L* suspended solids, and b) the percent volatile suspended solids of the sample.

	After Drying (Before Burning)	After Burning (Ash)
Weight of Sample & Dish	36.1544 g	36.1500 g
Weight of Dish (Tare Wt.)	36.1477 g	36.1477 g

Sample Volume = 25 m*L*

ANS_____

3. Given the following information regarding a treatment plant influent sample, calculate a) the mg/*L* suspended solids, and b) the percent volatile suspended solids of the sample.

	After Drying (Before Burning)	After Burning (Ash)
Weight of Sample & Dish	28.3169 g	28.3034 g
Weight of Dish (Tare Wt.)	28.2986 g	28.2986 g

Sample Volume = 25 m*L*

ANS_____

PRACTICE PROBLEMS 18.8: SVI and SDI

1. The settleability test indicates that after 30 minutes, 215 m*L* of settleable solids in the 1-liter graduated cylinder. If the Mixed Liquor Suspended Solids (MLSS) concentration in the aeration tank is 2180 mg/*L*, what is the sludge volume index?

ANS_____

2. The activated sludge settleability test indicates 380 m*L* settling in the 2-liter graduated cylinder. If the MLSS concentration in the aeration tank is 2260 mg/*L*, what is the sludge volume index?

ANS_____

3. The MLSS concentration in the aeration tank is 2050 mg/*L*. If the activated sludge settleability test indicates 219 m*L* settled in the one-liter graduated cylinder, what is the sludge density index?

ANS_____

4. The activated sludge in the aeration tank is found to have a 2050 mg/*L* MLSS concentration. If the settleability test indicates 182 m*L* settleable solids in one liter after 30 minutes, what is the sludge density index?

ANS_____

5. The settleability test indicates that after 30 minutes, 205 m*L* of solids settle in the 1-liter graduated cylinder. If the Mixed Liquor Suspended Solids (MLSS) concentration in the aerator is 2470 mg/*L*, what is the sludge volume index?

ANS_____

PRACTICE PROBLEMS 18.9: Temperature

1. The influent to a treatment plant has a temperature of 72° F. What is this temperature expressed in degrees Celsius?

ANS_____

2. Convert 56° Fahrenheit to degrees Celsius.

ANS_____

3. The effluent of a treatment plant is 22° C. What is this temperature expressed in degrees Fahrenheit?

ANS_____

4. What is 15° C expressed in terms of degrees Fahrenheit?

ANS_____

Chapter 18—Achievement Test

1. Calculate the percent removal of settleable solids if the settleable solids of the sedimentation tank influent are 18 mL/L and the settleable solids of the effluent are 0.9 mL/L

ANS_____

2. What is the molarity of a solution that has 0.75 moles solute dissolved in 1500 mL of solution?

ANS_____

3. The settleability test is conducted on a sample of MLSS. What is the percent settleable solids if 305 milliliters settle in the 2000-mL graduate?

ANS_____

4. If 2.2 equivalents of a chemical are dissolved in 1.25 liters solution, what is the normality of the solution? (Round to the nearest tenth.)

ANS_____

5. Given the following information, determine the BOD of the wastewater:

Sample Volume—5 m*L*
BOD Bottle Volume—300 m*L*
Initial DO of Diluted Sample—7.5 mg/*L*
DO of Diluted Sample—4.2 mg/*L*
(After 5 days)

ANS_____

6. Given the information below, determine (a) the percent total solids in the sample, and (b) the percent of volatile solids in the sludge sample:

	Sludge (Total Sample)	After Drying	After Burning (Ash)
Weight of Sample & Dish	80.14 g	27.70 g	26.14 g
Weight of Dish (tare wt.)	25.40 g	25.40 g	25.40 g

ANS_____

7. If magnesium has a listed atomic weight of 24, how many moles is represented by 80 grams of magnesium?

ANS_____

8. Given the following information regarding a treatment plant influent sample, calculate (a) the mg/*L* suspended solids, and (b) the percent volatile suspended solids of the sample.

	After Drying (Before Burning)	After Burning (Ash)
Weight of Sample & Dish	29.2686 g	29.2657 g
Weight of Dish (Tare Wt.)	29.2640 g	29.2640 g

Sample Volume = 50 m*L*

ANS_____

Chapter 18—Achievement Test—Cont'd

9. The settleability test indicates that after 30 minutes, 218 m*L* of sludge settle in the 1-liter graduated cylinder. If the Mixed Liquor Suspended Solids (MLSS) concentration in the aerator is 2310 mg/*L*, what is the sludge volume index?

ANS_____

10. The influent to a treatment plant has a temperature of 76°F. What is this temperature expressed in degrees Celsius?

ANS_____

11. Calculate the 7-day average primary effluent BOD for November 10th, 11th, and 12th, given the BOD test results shown below.

Nov. 1—205	Nov. 6 — 189	Nov. 11—212
Nov. 2—190	Nov. 7 — 196	Nov. 12—202
Nov. 3—192	Nov. 8 — 208	Nov. 13—218
Nov. 4—207	Nov. 9 — 211	Nov. 14—216
Nov. 5—194	Nov. 10—205	Nov. 15—222

ANS_____

12. The MLSS concentration in the aerator is 2190 mg/*L*. If the activated sludge settleability test indicates 190 m*L* settled in the one-liter graduated cylinder, what is the sludge density index?

ANS_____

13. How many milliliters of 0.5N NaOH will react with 800 m*L* of 0.1 N HCl?

ANS_____

14. Given the information before, calculate (a) the percent total solids and (b) percent volatile solids of the sludge sample.

	Sludge (Total Sample)	After Drying	After Burning (Ash)
Weight of Sample & Dish	75.02 g	21.92 g	19.78 g
Weight of Dish (tare wt.)	18.95 g	18.95 g	18.95 g

ANS_____

15. The influent to a treatment plant has a temperature of 18°C. What is this temperature expressed in degrees Fahrenheit?

ANS_____

16. The settleability test is conducted on a sample of MLSS. What is the percent settleable solids if 310 milliliters settle in the 2000-m*L* graduate?

ANS_____

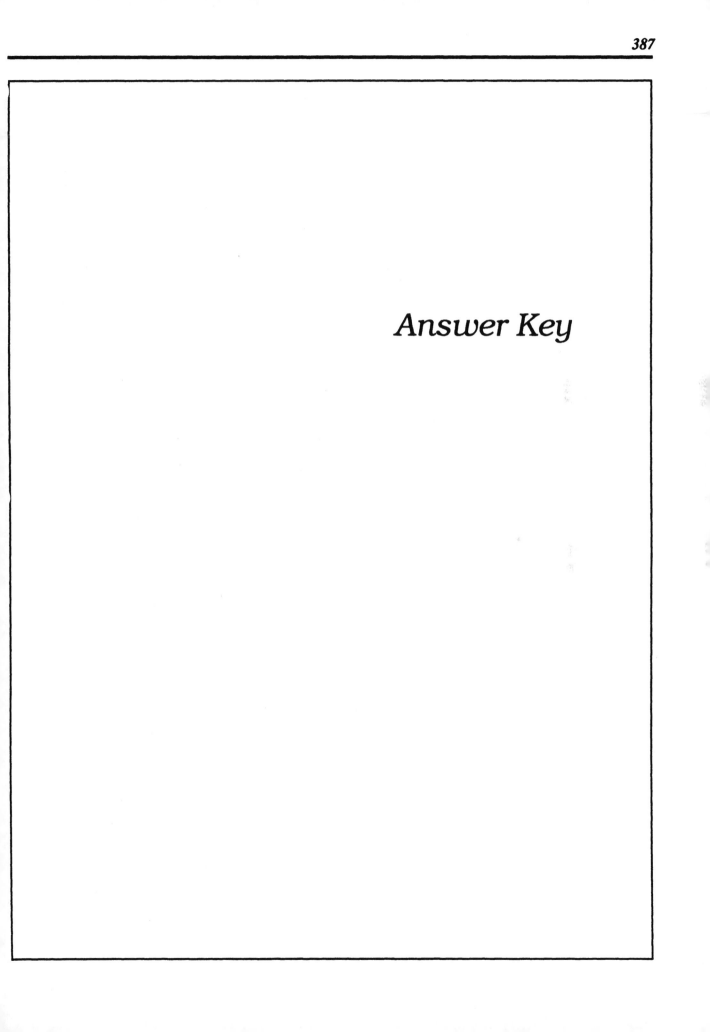

Answer Key

Wastewater Workbook
Answer Key

Chapter 1

PRACTICE PROBLEMS 1.1

1. $(0.785)(80 \text{ ft})(80 \text{ ft})(30 \text{ ft})(7.48 \text{ gal/cu ft}) = 1,127,386 \text{ gal}$

2. $(70 \text{ ft})(12 \text{ ft})(20 \text{ ft}) = 16,800 \text{ cu ft}$

3. $(25 \text{ ft})(80 \text{ ft})(13 \text{ ft})(7.48 \text{ gal/cu ft}) = 194,480 \text{ gal}$

4. $(15 \text{ ft})(30 \text{ ft})(10 \text{ ft})(7.48 \text{ gal/cu ft}) = 33,660 \text{ gal}$

5. $(0.785)(50 \text{ ft})(50 \text{ ft})(14 \text{ ft})(7.48 \text{ gal/cu ft}) = 205,513 \text{ gal}$

PRACTICE PROBLEMS 1.2

1. $(3 \text{ ft})(5 \text{ ft})(350 \text{ ft}) = 5250 \text{ cu ft}$

2. $(0.785)(0.83 \text{ ft})(0.83 \text{ ft})(1500 \text{ ft})(7.48 \text{ gal/cu ft}) = 6068 \text{ gal}$

3. $\dfrac{(5 \text{ ft} + 10 \text{ ft})}{2}(3 \text{ ft})(700 \text{ ft})(7.48 \text{ gal/cu ft}) = 117,810 \text{ gal}$

4. $(0.785)(0.5 \text{ ft})(0.5 \text{ ft})(1778 \text{ ft})(7.48 \text{ gal/cu ft}) = 2610 \text{ gal}$

5. $(5 \text{ ft})(3.7 \text{ ft})(1000 \text{ ft})(7.48 \text{ gal/cu ft}) = 138,380 \text{ gal}$

PRACTICE PROBLEMS 1.3

1. $\dfrac{(3 \text{ ft})(3.5 \text{ ft})(600 \text{ ft})}{27 \text{ cu ft/cu yd}} = 233 \text{ cu yds}$

2. $(430 \text{ ft})(660 \text{ ft})(6 \text{ ft}) = 1,702,800 \text{ cu ft}$

3. $(200 \text{ yds})(1 \text{ yd})(1.33 \text{ yds}) = 266 \text{ cu yds}$

4. $\dfrac{(5 \text{ ft} + 9 \text{ ft})}{2}(4.5 \text{ ft})(720 \text{ ft}) = 22,680 \text{ cu ft}$

5. $(750 \text{ ft})(2 \text{ ft})(2 \text{ ft}) = 3000 \text{ cu ft}$

CHAPTER 1 ACHIEVEMENT TEST

1. (600 ft)(3.5 ft)(6 ft) = 12,600 cu ft

2. (0.785)(70 ft)(70 ft)(23 ft)(7.48 gal/cu ft) = 661,752 gal

3. (430 ft)(660 ft)(4 ft) = 1,135,200 cu ft

4. (80 ft)(15 ft)(20 ft) = 24,000 cu ft

5. (0.785)(0.67 ft)(0.67 ft)(3500 ft)(7.48 gal/cu ft) = 9225 gal

6. (900 ft)(2 ft)(2.5 ft) = 4500 cu ft

7. $\dfrac{(2.5 \text{ ft})(3 \text{ ft})(1500 \text{ ft})}{27 \text{ cu ft/cu yd}} = 417$ cu yds

8. (5.5 ft)(2.5 ft)(1000 ft)(7.48 gal/cu ft) = 102,850 gal

9. (20 ft)(60 ft)(13 ft)(7.48 gal/cu ft) = 116,688 gal

10. (6 ft)(3.7 ft)(2000 ft)(7.48 gal/cu ft) = 332,112 gal

11. (7 ft)(3.5 ft)(851 ft) = 20,850 cu ft

12. (0.785)(50 ft)(50 ft)(16 ft)(7.48 gal/cu ft) = 234,872 gal

Chapter 2

PRACTICE PROBLEMS 2.1

1. (2.6 ft)(3.5 ft)(2.2 fps)(60 sec/min) = 1201 cfm

2. (15 ft)(10 ft)(0.7 fpm)(7.48 gal/cu ft) = 785 gpm

3. $\dfrac{(3.5 \text{ ft} + 5.5 \text{ ft})}{2}(3.2 \text{ ft})(125 \text{ fpm}) = 1800$ cfm

4. (0.785)(0.5 ft)(0.5 ft)(2.6 fps)(7.48 gal/cu ft)(60 sec/min) = 229 gpm

5. (0.785)(2 ft)(2 ft)(4.3 fpm)(7.48 gal/cu ft) = 101 gpm

6. (0.785)(0.83 ft)(0.83 ft)(3.2 fps)(7.48 gal/cu ft)(60 sec/min)(0.5) = 388 gpm

PRACTICE PROBLEMS 2.2

1. (5 ft)(2.3 ft)(x fps)(60 sec/min)(7.48 gal/cu ft) = 13,400 gpm
   x = 2.6 ft

2. (0.785)(0.67 ft)(0.67 ft)(x fps)(7.48 gal/cu ft)(60 sec/min) = 537 gpm
   x = 3.4 fps

3. 500 ft/208 sec = 2.4 fps

4. (0.785)(0.83 ft)(0.83 ft)(2.6 fps) = (0.785)(0.67 ft)(0.67 ft)(x fps)
   x = 4 fps

5. 400 ft/88 sec = 4.5 fps

6. (0.785)(0.67 ft)(0.67 ft)(3.6 fps) = (0.785)(0.83 ft)(0.83 ft)(x fps)
   x = 2.3 fps

PRACTICE PROBLEMS 2.3

1. 35.3 MGD/7 = 5 MGD

2. 117.3 MG/30 days = 3.9 MGD

3. 428.2 MG/91 days = 4.7 MGD

4. 3,140,000 gal/1440 min = 2180 gpm

PRACTICE PROBLEMS 2.4

1. (5 cfs)(7.48 gal/cu ft)(60 sec/min) = 2244 gpm

2. (38 gps)(60 sec/min)(1440 min/day) = 3,283,200 gpd

3. $\dfrac{4,270,000 \text{ gpd}}{(1440 \text{ min/day})(7.48 \text{ gal/cu ft})}$ = 396 cfm

4. (5.6 MGD)(1.55 cfs/MGD) = 8.7 cfs

5. $\dfrac{(423,690 \text{ cfd})(7.48 \text{ gal/cu ft})}{1440 \text{ min/day}}$ = 2201 gpm

6. (2730 gpm)(1440 min/day) = 3,931,200 gpd

CHAPTER 2 ACHIEVEMENT TEST

1. (6 ft)(2.6 ft)(*x* fps)(7.48 gal/cu ft)(60 sec/min) = 15,500 gpm
   x = 2.2 fps

2. 27.8 MGD/7 = 4 MGD

3. (4.2 ft)(3.2 ft)(3.9 fps)(60 sec/min) = 3145 cfm

4. 377.6 MG/92 days = 4.1 MGD

5. (10 ft)(10 ft)(0.67 fpm)(7.48 gal/cu ft) = 501 gpm

6. (0.785)(0.67 ft)(0.67 ft)(*x* fps)(7.48 gal/cu ft)(60 sec/min) = 490 gpm
   x = 3 fps

7. (8 cfs)(7.48 gal/cu ft)(60 sec/min) = 3590 gpm

8. 127.6 MG/31 days = 4.1 MGD

9. (4.8 MGD)(1.55 cfs/MGD) = 7.4 cfs

10. (0.785)(2 ft)(2 ft)(3 fpm)(7.48 gal/cu ft) = 70 gpm

11. $\dfrac{(1{,}780{,}000 \text{ gpd})}{(1440 \text{ min/day})(7.48 \text{ gal/cu ft})}$ = 165 cfm

12. (0.785)(0.5 ft)(0.5 ft)(2.7 fps)(7.48 gal/cu ft)(60 sec/min) = 238 gpm

13. 300 ft/86 sec = 3.5 fps

14. (0.785)(0.83 ft)(0.83 ft)(2.4 fps) = (0.785)(0.67 ft)(0.67 ft)(*x* fps)
    x = 3.7 fps

15. (2150 gpm)(1440 min/day) = 3,096,000 gpd

16. 4,620,000 gal/1440 min = 3208 gpm

Chapter 3

PRACTICE PROBLEMS 3.1

1. (2.3 mg/L)(5.1 MGD)(8.34 lbs/gal) = 98 lbs/day

2. (5.9 mg/L)(3.8 MGD)(8.34 lbs/gal) = 187 lbs/day

3. $\dfrac{(8 \text{ mg/L})(1.6 \text{ MGD})(8.34 \text{ lbs/gal})}{0.65}$ = 164 lbs/day

PRACTICE PROBLEMS 3.1—Cont'd

4. (10.5 mg/*L*)(4.6 MGD)(8.34 lbs/gal) = 403 lbs/day

5. (4.1 mg/*L*)(6.14 MGD)(8.34 lbs/gal) = 210 lbs/day

6. (50 mg/*L*)(0.085 MGD)(8.34 lbs/gal) = 35 lbs

7. (2120 mg/*L*)(0.224 MG)(8.34 lbs/gal) = 3960 lbs

8. $\dfrac{(8 \text{ mg/}L)(0.72 \text{ MGD})(8.34 \text{ lbs/gal})}{0.65}$ = 74 lbs/day

PRACTICE PROBLEMS 3.2

1. (425 mg/*L*)(1.62 MGD)(8.34 lbs/gal) = 5742 lbs/day

2. (26 mg/*L*)(2.98 MGD)(8.34 lbs/gal) = 646 lbs/day

3. (280 mg/*L*)(5.34 MGD)(8.34 lbs/gal) = 12,470 lbs/day

4. (135 mg/*L*)(3.54 MGD)(8.34 lbs/gal) = 3986 lbs/day

5. (295 mg/*L*)(3.30 MGD)(8.34 lbs/gal) = 8119 lbs/day

PRACTICE PROBLEMS 3.3

1. (148 mg/*L*)(5.2 MGD)(8.34 lbs/gal) = 6418 lbs/day SS

2. (189 mg/*L*)(1.89 MGD)(8.34 lbs/gal) = 2979 lbs/day SS

3. (136 mg/*L*)(4.79 MGD)(8.34 lbs/gal) = 5433 lbs/day BOD

4. (221 mg/*L*)(2.37 MGD)(8.34 lbs/gal) = 4368 solids

5. (118 mg/*L*)(4.14 MGD)(8.34 lbs/gal) = 4074 lbs/day

PRACTICE PROBLEMS 3.4

1. (2140 mg/*L*)(0.35 MG)(8.34 lbs/gal) = 6247 lbs SS

2. (1960 mg/*L*)(0.37 MG)(8.34 lbs/gal) = 6048 lbs MLVSS

3. (2960 mg/*L*)(0.18 MG)(8.34 lbs/gal) = 4444 lbs MLVSS

4. (2440 mg/*L*)(0.61 MG)(8.34 lbs/gal) = 12,413 lbs MLSS

5. (2890 mg/*L*)(0.49 MG)(8.34 lbs/gal) = 11,810 lbs MLSS

PRACTICE PROBLEMS 3.5

1. (6210 mg/L)(x MGD)(8.34 lbs/gal) = 5300 lbs/day
    x = 0.10 MGD

2. a) (5970 mg/L)(x MGD)(8.34 lbs/gal) = 4600 lbs/day
       x = 0.09 MGD

   b) 90,000 gpd/1440 min/day = 62.5 gpm

3. (6540 mg/L)(x MGD)(8.34 lbs/gal) = 6090 lbs/day
    x = 0.11 MGD

   Then 110,000 gpd ÷ 1440 min/day = 76.4 gpm

4. (6280 mg/L)(x MGD)(8.34 lbs/gal) = 7400 lbs/day
    x = 0.14 MGD

   Then 140,000 gpd ÷ 1440 min/day = 97 gpm

5. (7140 mg/L)(x MGD)(8.34 lbs/gal) = 5700 lbs/day
    x = 0.10 MGD

   Then 100,000 gpd ÷ 1440 min/day = 69 gpm

CHAPTER 3 ACHIEVEMENT TEST

1. (2.4 mg/L)(3.82 MGD)(8.34 lbs/gal) = 76 lbs/day

2. (15 mg/L)(2.05 MGD)(8.34 lbs/gal) = 256 lbs/day BOD

3. (185 mg/L)(4.6 MGD)(8.34 lbs/gal) = 7097 lbs/day SS Rem.

4. (9.9 mg/L)(5.6 MGD)(8.34 lbs/gal) = 462 lbs/day

5. (315 mg/L)(3.7 MGD)(8.34 lbs/gal) = 9720 lbs/day SS

6. $\dfrac{(12 \text{ mg/}L)(2.8 \text{ MGD})(8.34 \text{ lbs/gal})}{0.65}$ = 431 lbs/day Hypochlorite

7. (190 mg/L)(3.22 MGD)(8.34 lbs/gal) = 5102 lbs/day Solids

8. (50 mg/L)(0.08 MG)(8.34 lbs/gal) = 33 lbs Chlorine

9. (2610 mg/L)(0.39 MG)(8.34 lbs/gal) = 8489 lbs MLSS

10. (5980 mg/L)(x MGD)(8.34 lbs/gal) = 5540 lbs/day
     x = 0.11 MGD

CHAPTER 3 ACHIEVEMENT TEST—Cont'd

11. (125 mg/L)(3.168 MGD)(8.34 lbs/gal) = 3303 lbs/day COD

12. (235 mg/L)(3.15 MGD)(8.34 lbs/gal) = 6174 lbs BOD

13. (205 mg/L)(1.9 MGD)(8.34 lbs/gal) = 3248 lbs/day BOD Removed

14. (x mg/L)(5.1 MGD)(8.34 lbs/gal) = 340 lbs/day

    x = 8 mg/L

15. (5800 mg/L)(x MGD)(8.34 lbs/gal) = 6240 lbs/day

    x = 0.13 MGD

    Then, 130,000 gpd ÷ 1440 min/day = 90 gpm

Chapter 4

PRACTICE PROBLEMS 4.1

1. $$\frac{3,000,000 \text{ gpd}}{(0.785)(90 \text{ ft})(90 \text{ ft})} = 472 \text{ gpd/sq ft}$$

2. $$\frac{4,320,000 \text{ gpd}}{(0.785)(80 \text{ ft})(80 \text{ ft})} = 860 \text{ gpd/sq ft}$$

3. $$\frac{3,600,000 \text{ gpd}}{850,000 \text{ sq ft}} = 4.2 \text{ gpd/sq ft}$$

4. $$\frac{264,706 \text{ cu ft/day}}{653,400 \text{ sq ft}} = 0.4 \text{ ft/day}$$

    Then (0.4 ft/day)(12 in./ft) = 4.8 in./day

5. $$\frac{5,080,000 \text{ gpd}}{(0.785)(85 \text{ ft})(85 \text{ ft})} = 896 \text{ gpd/sq ft}$$

6. $$\frac{4.3 \text{ ac-ft/day}}{20 \text{ ac}} = 0.22 \text{ ft/day or 3 in./day}$$

PRACTICE PROBLEMS 4.2

1. $$\frac{2,140,000 \text{ gpd}}{(80 \text{ ft})(30 \text{ ft})} = 892 \text{ gpd/sq ft}$$

2. $\dfrac{2,620,000 \text{ gpd}}{(0.785)(70 \text{ ft})(70 \text{ ft})} = 681 \text{ gpd/sq ft}$

3. $\dfrac{3,280,000 \text{ gpd}}{(100 \text{ ft})(40 \text{ ft})} = 820 \text{ gpd/sq ft}$

4. $\dfrac{1,480,000 \text{ gpd}}{(20 \text{ ft})(80 \text{ ft})} = 925 \text{ gpd/sq ft}$

5. $\dfrac{2,360,000 \text{ gpd}}{(0.785)(60 \text{ ft})(60 \text{ ft})} = 835 \text{ gpd/sq ft}$

PRACTICE PROBLEMS 4.3

1. $\dfrac{2150 \text{ gpm}}{(30 \text{ ft})(25 \text{ ft})} = 2.9 \text{ gpm/sq ft}$

2. $\dfrac{3080 \text{ gpm}}{(40 \text{ ft})(20 \text{ ft})} = 3.9 \text{ gpm/sq ft}$

3. $\dfrac{2000 \text{ gpm}}{(25 \text{ ft})(50 \text{ ft})} = 1.6 \text{ gpm/sq ft}$

4. $\dfrac{1528 \text{ gpm}}{(45 \text{ ft})(25 \text{ ft})} = 1.4 \text{ gpm/sq ft}$

5. $\dfrac{2975 \text{ gpm}}{875 \text{ sq ft}} = 3.4 \text{ gpm/sq ft}$

PRACTICE PROBLEMS 4.4

1. $\dfrac{4950 \text{ gpm}}{(15 \text{ ft})(15 \text{ ft})} = 22 \text{ gpm/sq ft}$

2. $\dfrac{5100 \text{ gpm}}{(25 \text{ ft})(15 \text{ ft})} = 14 \text{ gpm/sq ft}$

3. $\dfrac{3300 \text{ gpm}}{(20 \text{ ft})(15 \text{ ft})} = 11 \text{ gpm/sq ft}$

4. $\dfrac{3200 \text{ gpm}}{(20 \text{ ft})(30 \text{ ft})} = 5.3 \text{ gpm/sq ft}$

5. $\dfrac{3800 \text{ gpm}}{(20 \text{ ft})(20 \text{ ft})} = 9.5 \text{ gpm/sq ft}$

PRACTICE PROBLEMS 4.5

1. $\dfrac{3{,}890{,}000 \text{ gal}}{(15 \text{ ft})(40 \text{ ft})} = 6483 \text{ gal/sq ft}$

2. $\dfrac{1{,}680{,}000 \text{ gpm}}{(15 \text{ ft})(15 \text{ ft})} = 7467 \text{ gal/sq ft}$

3. $\dfrac{3{,}960{,}000 \text{ gal}}{(20 \text{ ft})(25 \text{ ft})} = 7920 \text{ gal/sq ft}$

4. $\dfrac{1{,}339{,}200 \text{ gal}}{(15 \text{ ft})(12 \text{ ft})} = 7440 \text{ gal/sq ft}$

5. $\dfrac{5{,}625{,}000 \text{ gal}}{(30 \text{ ft})(25 \text{ ft})} = 7500 \text{ gal/sq ft}$

PRACTICE PROBLEMS 4.6

1. $\dfrac{1{,}397{,}000 \text{ gpd}}{157 \text{ ft}} = 8{,}898 \text{ gpd/ft}$

2. $\dfrac{2{,}320{,}000 \text{ gpd}}{(3.14)(60 \text{ ft})} = 12{,}314 \text{ gpd/ft}$

3. $\dfrac{2{,}600{,}000 \text{ gpd}}{235 \text{ ft}} = 11{,}064 \text{ gpd/ft}$

4. $\dfrac{(1200 \text{ gpm})(1440 \text{ min/day})}{(3.14)(70 \text{ ft})} = 7862 \text{ gpd/ft}$

5. $\dfrac{2819 \text{ gpm}}{188 \text{ ft}} = 15 \text{ gpm/ft}$

PRACTICE PROBLEMS 4.7

1. $\dfrac{(215 \text{ mg/L})(2.24 \text{ MGD})(8.34 \text{ lbs/gal})}{25.1 \ 1000\text{-cu ft}} = 160 \text{ lbs BOD/day/1000 cu ft}$

2. $\dfrac{(175 \text{ mg/L})(0.122 \text{ MGD})(8.34 \text{ lbs/gal})}{3.7 \text{ ac}} = 48 \text{ lbs BOD/day/ac}$

3. $\dfrac{(125 \text{ mg/L})(2.96 \text{ MGD})(8.34 \text{ lbs/gal})}{34 \ 1000\text{-cu ft}} = 91 \text{ lbs BOD/day/1000 cu ft}$

4. $\dfrac{(130 \text{ mg}/L)(2.15 \text{ MGD})(8.34 \text{ lbs/gal})}{800 \ 1000\text{-sq ft}}$ = 2.9 lbs BOD/day/1000 sq ft

5. $\dfrac{(140 \text{ mg}/L)(3.4 \text{ MGD})(8.34 \text{ lbs/gal})}{25.4 \ 1000\text{-cu ft}}$ = 156 lbs BOD/day/1000 cu ft

PRACTICE PROBLEMS 4.8

1. $\dfrac{(207 \text{ mg}/L)(3.35 \text{ MGD})(8.34 \text{ lbs/gal})}{(1950 \text{ mg}/L)(0.4 \text{ MG})(8.34 \text{ lbs/gal})}$ = 0.9

2. $\dfrac{(192 \text{ mg}/L)(3.15 \text{ MGD})(8.34 \text{ lbs/gal})}{(1690 \text{ mg}/L)(0.27 \text{ MG})(8.34 \text{ lbs/gal})}$ = 1.3

3. $\dfrac{(147 \text{ mg}/L)(2.26 \text{ MGD})(8.34 \text{ lbs/gal})}{x \text{ lbs MLVSS}}$ = 0.7

   x = 3958 lbs MLVSS

4. $\dfrac{(155 \text{ mg}/L)(1.92 \text{ MGD})(8.34 \text{ lbs/gal})}{(1880 \text{ mg}/L)(0.245 \text{ MG})(8.34 \text{ lbs/gal})}$ = 0.6

5. $\dfrac{(185 \text{ mg}/L)(2.94 \text{ MGD})(8.34 \text{ lbs/gal})}{(x \text{ mg}/L)(0.45 \text{ MG})(8.34 \text{ lbs/gal}}$ = 0.5

   x = 2417 mg/L MLVSS

PRACTICE PROBLEMS 4.9

1. $\dfrac{(2780 \text{ mg}/L)(3.75 \text{ MGD})(8.34 \text{ lbs/gal})}{(0.785)(80 \text{ ft})(80 \text{ ft})}$ = 17.3 lbs MLSS/day/sq ft

2. $\dfrac{(2950 \text{ mg}/L)(4.07 \text{ MGD})(8.34 \text{ lbs/gal})}{(0.785)(75 \text{ ft})(75 \text{ ft})}$ = 22.7 lbs MLSS/day/sq ft

3. $\dfrac{(x \text{ mg}/L)(3.59 \text{ MGD})(8.34 \text{ lbs/gal})}{(0.785)(60 \text{ ft})(60 \text{ ft})}$ = 28 lbs MLSS/day/sq ft

   x = 2643 mg/L MLSS

4. $\dfrac{(2180 \text{ mg}/L)(3.2 \text{ MGD})(8.34 \text{ lbs/gal})}{(0.785)(50 \text{ ft})(50 \text{ ft})}$ = 29.6 lbs MLSS/day/sq ft

PRACTICE PROBLEMS 4.9—Cont'd

5. $\dfrac{(x \text{ mg}/L)(2.96 \text{ MGD})(8.34 \text{ lbs/gal})}{(0.785)(55 \text{ ft})(55 \text{ ft})} = 20 \text{ lbs MLSS/day/sq ft}$

  $x = 1924 \text{ mg}/L \text{ MLSS}$

PRACTICE PROBLEMS 4.10

1. $\dfrac{11,650 \text{ lbs VS/day}}{32,600 \text{ cu ft}} = 0.36 \text{ lbs VS/day/cu ft}$

2. $\dfrac{(122,000 \text{ lbs/day})(0.065)(0.70)}{(0.785)(50 \text{ ft})(50 \text{ ft})(20 \text{ ft})} = 0.14 \text{ lbs VS/day/cu ft}$

3. $\dfrac{(139,000 \text{ lbs/day})(0.06)(0.69)}{(0.785)(45 \text{ ft})(45 \text{ ft})(19 \text{ ft})} = 0.19 \text{ lbs VS/day/cu ft}$

4. $\dfrac{(19,200 \text{ gpd})(8.34 \text{ lbs/gal})(0.053)(0.68)}{(0.785)(35 \text{ ft})(35 \text{ ft})(15 \text{ ft})} = 0.40 \text{ lbs VS/day/cu ft}$

5. $\dfrac{(21,000 \text{ gpd})(8.8 \text{ lbs/gal})(0.052)(0.71)}{(0.785)(40 \text{ ft})(40 \text{ ft})(19 \text{ ft})} = 0.29 \text{ lbs VS/day/cu ft}$

PRACTICE PROBLEMS 4.11

1. $\dfrac{1930 \text{ lbs VS Added/day}}{31,200 \text{ lbs VS}} = 0.06$

2. $\dfrac{550 \text{ lbs VS Added/day}}{(172,700 \text{ lbs})(0.058)(0.65)} = 0.08$

3. $\dfrac{(61,200 \text{ lbs/day})(0.054)(0.71)}{(110,000 \text{ gal})(8.34 \text{ lbs/gal})(0.065)(0.57)} = 0.07$

4. $\dfrac{x \text{ lbs VS Added/day}}{(108,000 \text{ gal})(8.34 \text{ lbs/gal})(0.058)(0.56)} = 0.08$

  $x = 2340 \text{ lbs/day VS}$

5. $\dfrac{(7,700 \text{ gpd})(8.34 \text{ lbs/day})(0.044)(0.72)}{x \text{ lbs VS}} = 0.06$

  $x = 33,907 \text{ lbs VS}$

PRACTICE PROBLEMS 4.12

1. 1530 people/4.7 ac = 326 people/ac

2. 3825 people/9 ac = 425 people/ac

3. $\dfrac{(1710 \text{ mg}/L)(0.372 \text{ MGD})(8.34 \text{ lbs/gal})}{0.2 \text{ lbs/day}}$ = 26,526 people

4. $\dfrac{5000 \text{ people}}{x \text{ ac}}$ = 350 people/ac

   $x = 14.3$ ac

5. $\dfrac{(2190 \text{ mg}/L)(0.098 \text{ MGD})(8.34 \text{ lbs/gal})}{0.2 \text{ lbs/day}}$ = 8950 people

CHAPTER 4 ACHIEVEMENT TEST

1. $\dfrac{2,140,000 \text{ gpd}}{(0.785)(75 \text{ ft})(75 \text{ ft})}$ = 485 gpd/sq ft

2. $\dfrac{2940 \text{ gpm}}{180 \text{ sq ft}}$ = 16.3 gpm/sq ft

3. $\dfrac{1,990,000 \text{ gpd}}{(3.14)(75 \text{ ft})}$ = 8450 gpd/ft

4. $\dfrac{3,100,000 \text{ gpd}}{(0.785)(80 \text{ ft})(80 \text{ ft})}$ = 617 gpd/sq ft

5. $\dfrac{(156 \text{ mg}/L)(1.8 \text{ MGD})(8.34 \text{ lbs/gal})}{x \text{ lbs MLVSS}}$ = 0.6

   $x = 3903$ lbs MLVSS

6. $\dfrac{400 \text{ lbs VS Added/day}}{(175,000 \text{ lbs})(0.062)(0.68)}$ = 0.05

7. $\dfrac{(2640 \text{ mg}/L)(3.45 \text{ MGD})(8.34 \text{ lbs/gal})}{(0.785)(75 \text{ ft})(75 \text{ ft})}$ = 17 lbs/day/sq ft

8. $\dfrac{(110,000 \text{ lbs/day})(0.068)(0.70)}{(0.785)(60 \text{ ft})(60 \text{ ft})(22 \text{ ft})}$ = 0.08

CHAPTER 4 ACHIEVEMENT TEST—Cont'd

9. $\dfrac{3.95 \text{ ac-ft/day}}{20 \text{ ac}} = 0.2 \text{ ft/day}$

   Then $(0.2 \text{ ft/day})(12 \text{ in./ft}) = 2.4 \text{ in./day}$

10. $\dfrac{(172 \text{ mg/}L)(3.154 \text{ MGD})(8.34 \text{ lbs/gal})}{(x \text{ mg/}L)(0.377 \text{ MG})(8.34 \text{ lbs/gal})} = 0.5$

    $x = 2878 \text{ mg/}L \text{ MLVSS}$

11. $\dfrac{(3180 \text{ mg/}L)(4.1 \text{ MGD})(8.34 \text{ lbs/gal})}{(0.785)(70 \text{ ft})(70 \text{ ft})} = 28 \text{ lbs/day/sq ft}$

12. $\dfrac{2,100,000 \text{ gpd}}{(90 \text{ ft})(30 \text{ ft})} = 778 \text{ gpd/sq ft}$

13. $\dfrac{1,740,000 \text{ gal}}{(20 \text{ ft})(15 \text{ ft})} = 5800 \text{ gal/sq ft}$

14. $\dfrac{(140 \text{ mg/}L)(2.68 \text{ MGD})(8.34 \text{ lbs/gal})}{(1890 \text{ mg/}L)(0.29 \text{ MG})(8.34 \text{ lbs/gal})} = 0.7$

15. $\dfrac{x \text{ lbs VS Added/day}}{(23,800 \text{ gal})(8.34 \text{ lbs/gal})(0.057)(0.58)} = 0.07$

    $x = 459 \text{ lbs/day}$

16. $\dfrac{3000 \text{ gpm}}{(40 \text{ ft})(25 \text{ ft})} = 3 \text{ gpm/sq ft}$

17. $\dfrac{(110 \text{ mg/}L)(3.1 \text{ MGD})(8.34 \text{ lbs/gal})}{20.1 \quad 1000\text{-cu ft}} = 141 \text{ lbs BOD/day/1000 cu ft}$

18. $\dfrac{2,460,000 \text{ gal}}{(3.14)(70 \text{ ft})} = 11,192 \text{ gpd/ft}$

19. $1800 \text{ people/}5.2 \text{ ac} = 346 \text{ people/ac}$

20. $\dfrac{(120 \text{ mg/}L)(2.41 \text{ MGD})(8.34)}{750 \quad 1000\text{-sq ft}} = 3.2 \text{ lbs BOD/day/1000 sq ft}$

21. $\dfrac{1889 \text{ gpm}}{(40 \text{ ft})(20 \text{ ft})} = 2.4 \text{ gpm/sq ft}$

Chapter 5

PRACTICE PROBLEMS 5.1

1. $\dfrac{(30 \text{ ft})(15 \text{ ft})(7 \text{ ft})(7.48 \text{ gal/cu ft})}{937.5 \text{ gpm}} = 25 \text{ min}$

2. $\dfrac{(75 \text{ ft})(25 \text{ ft})(10 \text{ ft})(7.48 \text{ gal/cu ft})}{66,667 \text{ gph}} = 2.1 \text{ hrs}$

3. $\dfrac{(4 \text{ ft})(5 \text{ ft})(2.5 \text{ ft})(7.48 \text{ gal/cu ft})}{(5 \text{ gpm})(60 \text{ min/hr})} = 1.2 \text{ hrs}$

4. $\dfrac{(0.785)(70 \text{ ft})(70 \text{ ft})(12 \text{ ft})(7.48 \text{ gal/cu ft})}{202,917 \text{ gph}} = 1.7 \text{ hrs}$

5. $\dfrac{(400 \text{ ft})(550 \text{ ft})(5 \text{ ft})(7.48 \text{ gal/cu ft})}{219,400 \text{ gpd}} = 37.5 \text{ days}$

PRACTICE PROBLEMS 5.2

1. $\dfrac{11,900 \text{ lbs MLSS}}{2627 \text{ lbs/day SS}} = 4.5 \text{ days}$

2. $\dfrac{(2670 \text{ mg/}L \text{ MLSS})(0.28 \text{ MG})(8.34 \text{ lbs/gal})}{(128 \text{ mg/}L)(0.958 \text{ MGD})(8.34 \text{ lbs/gal})} = 6.1 \text{ days}$

3. $\dfrac{(2960 \text{ mg/}L \text{ MLSS})(0.22 \text{ MG})(8.34 \text{ lbs/gal})}{(84 \text{ mg/}L)(1.46 \text{ MGD})(8.34 \text{ lbs/gal})} = 5.3 \text{ days}$

4. $\dfrac{(x \text{ mg/}L \text{ MLSS})(0.195 \text{ MG})(8.34 \text{ lbs/gal})}{(70 \text{ mg/}L)(1.25 \text{ MGD})(8.34 \text{ lbs/gal})} = 6.5 \text{ days}$

   $x = 2917 \text{ mg/}L \text{ MLSS}$

5. $\dfrac{x \text{ lbs MLSS}}{1560 \text{ lbs/day SS}} = 5.5 \text{ days}$

   $x = 8580 \text{ lbs MLSS}$

PRACTICE PROBLEMS 5.3

1. $\dfrac{(3100 \text{ mg/}L)(0.46 \text{ MG})(8.34 \text{ lbs/gal})}{1540 \text{ lbs/day Wasted} + 330 \text{ lbs/day in SE}} = 6.4 \text{ days}$

2. $\dfrac{(2810 \text{ mg/}L \text{ MLSS})(0.325 \text{ MG})(8.34 \text{ lbs/gal})}{(5340 \text{ mg/}L \text{ SS})(0.0185 \text{ MG})(8.34 \text{ lbs/gal}) + (15 \text{ mg/}L \text{ SS})(2.15 \text{ MGD})(8.34 \text{ lbs/gal})}$

$= \dfrac{7617 \text{ lbs}}{824 \text{ lbs/day} + 269 \text{ lbs/day}} = 7.0 \text{ days}$

3. $\dfrac{(2440 \text{ mg/}L \text{ MLSS})(1.5 \text{ MG})(8.34 \text{ lbs/gal})}{(6120 \text{ mg/}L \text{ SS})(0.075 \text{ MGD})(8.34 \text{ lbs/gal}) + (18 \text{ mg/}L)(2.6 \text{ MGD})(8.34 \text{ lbs/gal})}$

$= \dfrac{30{,}524 \text{ lbs MLSS}}{3828 \text{ lbs/day} + 390 \text{ lbs/day}} = 7.2 \text{ days}$

4. $\dfrac{(x \text{ mg/}L)(0.755 \text{ MG})(8.34 \text{ lbs/gal})}{(6320 \text{ mg/}L)(0.030 \text{ MGD})(8.34 \text{ lbs/gal}) + (22 \text{ mg/}L)(2.4 \text{ MGD})(8.34 \text{ lbs/gal})} = 8 \text{ days}$

First simplify:

$\dfrac{(x \text{ mg/}L)(0.755 \text{ MG})(8.34 \text{ lbs/gal})}{1581 \text{ lbs/day} + 440 \text{ lbs/day}} = 8 \text{ days}$

$\dfrac{(x \text{ mg/}L)(0.755 \text{ MG})(8.34 \text{ lbs/gal})}{2021 \text{ lbs/day}} = 8 \text{ days}$

Now solve for the unknown value:

$x = 2568 \text{ mg/}L \text{ MLSS}$

CHAPTER 5 ACHIEVEMENT TEST

1. $\dfrac{(70 \text{ ft})(25 \text{ ft})(12 \text{ ft})(7.48 \text{ gal/cu ft})}{65{,}833 \text{ gph}} = 2.4 \text{ hrs}$

2. $\dfrac{12{,}400 \text{ lbs MLSS}}{2750 \text{ lbs/day SS}} = 4.5 \text{ days}$

3. $\dfrac{(2950 \text{ mg/}L \text{ MLSS})(0.47 \text{ MG})(8.34 \text{ lbs/gal})}{1620 \text{ lbs/day wasted} + 310 \text{ lbs/day in SE}} = 6.0 \text{ days}$

4. $\dfrac{(35 \text{ ft})(15 \text{ ft})(8 \text{ ft})(7.48 \text{ gal/cu ft})}{1125 \text{ gpm}} = 28 \text{ min}$

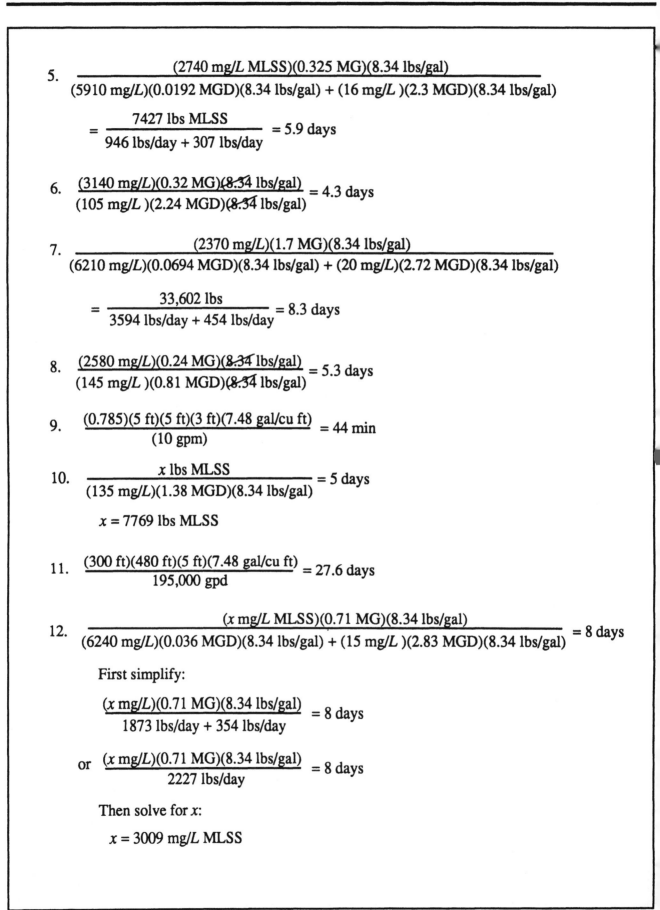

5.
$$\frac{(2740 \text{ mg}/L \text{ MLSS})(0.325 \text{ MG})(8.34 \text{ lbs/gal})}{(5910 \text{ mg}/L)(0.0192 \text{ MGD})(8.34 \text{ lbs/gal}) + (16 \text{ mg}/L)(2.3 \text{ MGD})(8.34 \text{ lbs/gal})}$$

$$= \frac{7427 \text{ lbs MLSS}}{946 \text{ lbs/day} + 307 \text{ lbs/day}} = 5.9 \text{ days}$$

6.
$$\frac{(3140 \text{ mg}/L)(0.32 \text{ MG})(8.34 \text{ lbs/gal})}{(105 \text{ mg}/L)(2.24 \text{ MGD})(8.34 \text{ lbs/gal})} = 4.3 \text{ days}$$

7.
$$\frac{(2370 \text{ mg}/L)(1.7 \text{ MG})(8.34 \text{ lbs/gal})}{(6210 \text{ mg}/L)(0.0694 \text{ MGD})(8.34 \text{ lbs/gal}) + (20 \text{ mg}/L)(2.72 \text{ MGD})(8.34 \text{ lbs/gal})}$$

$$= \frac{33,602 \text{ lbs}}{3594 \text{ lbs/day} + 454 \text{ lbs/day}} = 8.3 \text{ days}$$

8.
$$\frac{(2580 \text{ mg}/L)(0.24 \text{ MG})(8.34 \text{ lbs/gal})}{(145 \text{ mg}/L)(0.81 \text{ MGD})(8.34 \text{ lbs/gal})} = 5.3 \text{ days}$$

9.
$$\frac{(0.785)(5 \text{ ft})(5 \text{ ft})(3 \text{ ft})(7.48 \text{ gal/cu ft})}{(10 \text{ gpm})} = 44 \text{ min}$$

10.
$$\frac{x \text{ lbs MLSS}}{(135 \text{ mg}/L)(1.38 \text{ MGD})(8.34 \text{ lbs/gal})} = 5 \text{ days}$$

$$x = 7769 \text{ lbs MLSS}$$

11.
$$\frac{(300 \text{ ft})(480 \text{ ft})(5 \text{ ft})(7.48 \text{ gal/cu ft})}{195,000 \text{ gpd}} = 27.6 \text{ days}$$

12.
$$\frac{(x \text{ mg}/L \text{ MLSS})(0.71 \text{ MG})(8.34 \text{ lbs/gal})}{(6240 \text{ mg}/L)(0.036 \text{ MGD})(8.34 \text{ lbs/gal}) + (15 \text{ mg}/L)(2.83 \text{ MGD})(8.34 \text{ lbs/gal})} = 8 \text{ days}$$

First simplify:

$$\frac{(x \text{ mg}/L)(0.71 \text{ MG})(8.34 \text{ lbs/gal})}{1873 \text{ lbs/day} + 354 \text{ lbs/day}} = 8 \text{ days}$$

or
$$\frac{(x \text{ mg}/L)(0.71 \text{ MG})(8.34 \text{ lbs/gal})}{2227 \text{ lbs/day}} = 8 \text{ days}$$

Then solve for x:

$$x = 3009 \text{ mg}/L \text{ MLSS}$$

Chapter 6

PRACTICE PROBLEMS 6.1

1. $\dfrac{98 \text{ mg/}L \text{ Rem.}}{120 \text{ mg/}L} \times 100 = 82\%$

2. $\dfrac{230 \text{ mg/}L \text{ Rem.}}{245 \text{ mg/}L} \times 100 = 94\%$

3. $\dfrac{208 \text{ mg/}L \text{ Rem.}}{270 \text{ mg/}L} \times 100 = 77\%$

4. $\dfrac{95 \text{ mg/}L \text{ Rem.}}{230 \text{ mg/}L} \times 100 = 41\%$

5. $\dfrac{168 \text{ mg/}L \text{ Rem.}}{305 \text{ mg/}L} \times 100 = 55\%$

PRACTICE PROBLEMS 6.2

1. $5.8 = \dfrac{x \text{ lbs/day Solids}}{(3610 \text{ gal})(8.34 \text{ lbs/gal})} \times 100$

   $x = 1746$ lbs/day Solids

2. $\dfrac{0.68 \text{ grams Solids}}{13.05 \text{ grams Sludge}} \times 100 = 5.2\%$ solids

3. $\dfrac{1430 \text{ lbs/day Solids}}{x \text{ lbs/day Sludge}} \times 100 = 3.2$

   $x = 44,688$ lbs/day

4. $4.4 = \dfrac{265 \text{ lbs/day SS}}{(x \text{ gpd})(8.34 \text{ lbs/gal})} \times 100$

   $x = 722$ gpd

5. $3.4 = \dfrac{x \text{ lbs/day Solids}}{286,000 \text{ lbs/day Sludge}} \times 100$

   $x = 9724$ lbs/day Solids

PRACTICE PROBLEMS 6.3

1. $$\frac{\frac{(3000 \text{ gpd})(8.34 \text{ lbs/gal})(4.5)}{100} + \frac{(4200 \text{ gpd})(8.34 \text{ lbs/gal})(3.8)}{100}}{(3000 \text{ gpd})(8.34 \text{ lbs/gal}) + (4200 \text{ gpd})(8.34 \text{ lbs/gal})} \times 100$$

First simplify:

$$\frac{1126 \text{ lbs/day Sol.} + 1331 \text{ lbs/day Sol.}}{25,020 \text{ lbs/day} + 35,028 \text{ lbs/day sludge}} \times 100$$
$$\text{sludge}$$

$$= \frac{2457 \text{ lbs/day Solids}}{60,048 \text{ lbs/day Sludge}} \times 100$$

$$= 4.1\%$$

2. $$\frac{\frac{(8200 \text{ gpd})(8.34 \text{ lbs/gal})(5.2)}{100} + \frac{(7000 \text{ gpd})(8.34 \text{ lbs/gal})(4.2)}{100}}{(8200 \text{ gpd})(8.34 \text{ lbs/gal}) + (7000 \text{ gpd})(8.34 \text{ lbs/gal})} \times 100$$

Simplify:

$$\frac{3556 \text{ lbs/day Sol.} + 2452 \text{ lbs/day Sol.}}{68,388 \text{ lbs/day Sludge} + 58,380 \text{ lbs/day Sludge}} \times 100$$

$$= \frac{6008 \text{ lbs/day Solids}}{126,768 \text{ lbs/day Sludge}} \times 100$$

$$= 4.7\%$$

3. $$\frac{\frac{(4840 \text{ gpd})(8.34)(4.9)}{100} + \frac{(5200 \text{ gpd})(8.34)(3.6)}{100}}{(4840)(8.34) + (5200)(8.34)} \times 100$$

$$= \frac{1978 \text{ lbs/day} + 1561 \text{ lbs/day}}{40,366 \text{ lbs/day} + 43,368 \text{ lbs/day}} \times 100$$

$$= \frac{3539 \text{ lbs/day Solids}}{83,734 \text{ lbs/day Sludge}} \times 100$$

$$= 4.2\%$$

PRACTICE PROBLEMS 6.3—Cont'd

4. $$\dfrac{(9010\ \text{gpd})(8.34)\dfrac{(4.2)}{100} + (10{,}760\ \text{gpd})(8.34)\dfrac{(6.5)}{100}}{(9010)(8.34) + (10{,}760)(8.34)} \times 100$$

$$= \dfrac{3156\ \text{lbs/day} + 5833\ \text{lbs/day}}{75{,}143\ \text{lbs/day} + 89{,}738\ \text{lbs/day}} \times 100$$

$$= \dfrac{8989\ \text{lbs/day}}{164{,}881\ \text{lbs/day}} \times 100$$

$$= 5.5\%$$

PRACTICE PROBLEMS 6.4

1. (3340 lbs/day Solids)(0.70) = 2338 lbs/day VS

2. (4070 gpd Sludge)(8.34 lbs/gal)(0.07)(0.72) = 1711 lbs/day VS

3. (6400 gpd Sludge)(8.34 lbs/gal)(0.065)(0.67) = 2325 lbs/day VS

4. (24,510 lbs/day Sludge)(0.068)(0.66) = 1100 lbs/day VS

5. (2530 gpd Sludge)(8.34 lbs/gal)(0.062)(0.71) = 929 lbs/day VS

PRACTICE PROBLEMS 6.5

1. $$18 = \dfrac{x\ \text{gal Seed Sludge}}{280{,}000\ \text{gal Volume}} \times 100$$

    x = 50,400 gal Seed Sludge

2. $$20 = \dfrac{x\ \text{gal Seed Sludge}}{(0.785)(70\ \text{ft})(70\ \text{ft})(21\ \text{ft})(7.48\ \text{gal/cu ft})} \times 100$$

    x = 120,842 gal Seed Sludge

3. $$15 = \dfrac{x\ \text{gal Seed Sludge}}{(0.785)(50\ \text{ft})(50\ \text{ft})(21\ \text{ft})(7.48\ \text{gal/cu ft})} \times 100$$

    x = 46,240 gal Seed Sludge

4. $$x = \dfrac{87{,}500\ \text{gal}}{(0.785)(60\ \text{ft})(60\ \text{ft})(23\ \text{ft})(7.48\ \text{gal/cu ft})} \times 100$$

    x = 18%

PRACTICE PROBLEMS 6.6

1. $x = \dfrac{2 \text{ lbs Chemical}}{80 \text{ lbs Solution}} \times 100$

   $x = 2.5\%$ Strength

2. 6 oz Polymer = 0.4 lbs Polymer)

   $x = \dfrac{0.4 \text{ lbs Polymer}}{(8 \text{ gal})(8.34 \text{ lbs/gal}) + 0.4 \text{ lbs}} \times 100$

   $x = \dfrac{.4}{66.7 \text{ lbs} + 0.4 \text{ lbs}} \times 100$

   $x = 0.6\%$ Strength

3. $1.5 = \dfrac{x \text{ lbs Polymer}}{(50 \text{ gal})(8.34 \text{ lbs/gal}) + x \text{ lbs Polymer}} \times 100$

   $0.015 = \dfrac{x}{417 + x}$

   $(0.015)(417 + x) = x$

   $6.255 + 0.015\,x = x$

   $6.255 = x - 0.015x$

   $6.255 = 0.985x$

   $6.4 \text{ lbs} = x$

4. (500 g = 1.1 lbs dry polymer)

   $x = \dfrac{1.1 \text{ lbs Polymer}}{(8 \text{ gal})(8.34 \text{ lbs/gal}) + 1.1 \text{ lbs}} \times 100$

   $x = \dfrac{1.1}{66.7 + 1.1} \times 100$

   $x = 1.6\%$

PRACTICE PROBLEMS 6.6—Cont'd

5. First calculate lbs of chemical required:

$$1.8 = \frac{x \text{ lbs Chemical}}{(5 \text{ gal})(8.34 \text{ lbs/gal}) + x \text{ lbs}} \times 100$$

$$\frac{1.8}{100} = \frac{x \text{ lbs}}{41.7 \text{ lbs} + x \text{ lbs}}$$

$$0.018 = \frac{x}{41.7 \text{ lbs} + x \text{ lbs}}$$

$$(0.018) \ (41.7 + x) = x$$

$$0.75 + 0.018\,x = x$$

$$0.75 = x - 0.018\,x$$

$$0.75 = 0.982\,x$$

$$0.76 \text{ lbs} = x$$

Then determine grams chem. required:

$$(0.76 \text{ lbs})(454 \text{ grams/lb}) = 345 \text{ grams}$$

PRACTICE PROBLEMS 6.7

1.
$$x = \frac{\dfrac{(15 \text{ lbs Sol'n 1})(9)}{100} + \dfrac{(100 \text{ lbs Sol'n 2})(1)}{100}}{15 \text{ lbs} + 100 \text{ lbs}} \times 100$$

$$x = \frac{(15)(0.09) + (100)(0.01)}{115} \times 100$$

$$x = \frac{1.35 + 1}{115} \times 100$$

$$x = 2\% \text{ Strength}$$

2.
$$x = \frac{\dfrac{(10 \text{ lbs})(12)}{100} + \dfrac{(330 \text{ lbs})(0.3)}{100}}{10 \text{ lbs} + 330 \text{ lbs}} \times 100$$

$$x = \frac{(10)(0.12) + (330)(0.003)}{340} \times 100$$

$$x = \frac{1.2 + 0.99}{340} \times 100$$

$$x = 0.6\% \text{ Strength}$$

3.
$$x = \frac{\dfrac{(20\ lbs)(12)}{100} + \dfrac{(445\ lbs)(0.5)}{100}}{20\ lbs + 445\ lbs} \times 100$$

$$x = \frac{(20)(0.12) + (445)(0.005)}{465} \times 100$$

$$x = \frac{2.4\ lbs + 2.2\ lbs}{465\ lbs} \times 100$$

$$x = 1.0\% \text{ Strength}$$

4. First determine lbs of each solution:

11% solution: (10 gal)(9.8 lbs/gal) = 98 lbs

0.1% solution: (60 gal)(8.34 lbs/gal) = 500 lbs

Then continue as with problems 1-3:

$$x = \frac{\dfrac{(98\ lbs)(11)}{100} + \dfrac{(500\ lbs)(0.1)}{100}}{98\ lbs + 500\ lbs} \times 100$$

$$x = \frac{10.8\ lbs + 0.5\ lbs}{598\ lbs} \times 100$$

$$x = 1.9\% \text{ Strength}$$

PRACTICE PROBLEMS 6.8

1. $x = \dfrac{17\ hp}{22\ hp} \times 100$

$x = 77\%$

2. $x = \dfrac{43\ hp}{50\ hp} \times 100$

$x = 86\%$

3. $89 = \dfrac{x}{25} \times 100$

$x = 22\ hp$

4. $62 = \dfrac{x}{50} \times 100$

$x = 31\ hp$

PRACTICE PROBLEMS 6.8—Cont'd

5. $86 = \dfrac{34.4 \text{ hp}}{x} \times 100$

   $x = 40 \text{ hp}$

CHAPTER 6 ACHIEVEMENT TEST

1. $x = \dfrac{209}{230} \times 100$

   $x = 91\%$

2. $(7200 \text{ gpd})(8.34 \text{ lbs/gal})(0.07)(0.68) = 2858 \text{ lbs/day VS}$

3. $x = \dfrac{(2600 \text{ gpd})(8.34 \text{ lbs/gal})\dfrac{(4.2)}{100} + (3380 \text{ gpd})(8.34 \text{ lb/gal})\dfrac{(3.5)}{100}}{(2600 \text{ gpd})(8.34 \text{ lbs/gal}) + (3380 \text{ gpd})(8.34 \text{ lb/gal})} \times 100$

   $x = \dfrac{911 \text{ lbs/day} + 987 \text{ lbs/day}}{21,684 \text{ lbs/day} + 28,189 \text{ lbs/day}} \times 100$

   $x = \dfrac{1898 \text{ lbs/day}}{49,873 \text{ lbs/day}} \times 100$

   $x = 4\%$

4. $x = \dfrac{3 \text{ lbs Chemical}}{92 \text{ lbs Solution}} \times 100$

   $x = 3\%$

5. $20 = \dfrac{x \text{ gal}}{(0.785)(80 \text{ ft})(80 \text{ ft})(23 \text{ ft})(7.48 \text{ gal/cu ft})} \times 100$

   $x = 172,866 \text{ gal}$

6. $x = \dfrac{(12 \text{ lbs})\dfrac{(10)}{100} + (400 \text{ lbs})\dfrac{(0.15)}{100}}{12 \text{ lbs} + 400 \text{ lbs}} \times 100$

   $x = \dfrac{1.2 \text{ lbs} + 0.6 \text{ lbs}}{412 \text{ lbs}} \times 100$

   $x = 0.4\%$

7. $x = \dfrac{20.4 \text{ hp}}{25.5 \text{ hp}} \times 100$

   $x = 80\%$ Effic.

8. $x = \dfrac{269 \text{ mg/L}}{350 \text{ mg/L}} \times 100$

   $x = 77\%$ Rem. Effic.

9. $x = \dfrac{\dfrac{(8700 \text{ gpd})(8.34 \text{ lbs/gal})(4.8)}{100} + \dfrac{(11,200 \text{ gpd})(8.34 \text{ lb/gal})(6.2)}{100}}{(8700 \text{ gpd})(8.34 \text{ lbs/gal}) + (11,200 \text{ gpd})(8.34 \text{ lb/gal})} \times 100$

   $x = \dfrac{3483 \text{ lbs/day} + 5791 \text{ lbs/day}}{72,558 \text{ lbs/day} + 93,408 \text{ lbs/day}} \times 100$

   $x = \dfrac{9274 \text{ lbs/day}}{165,966 \text{ lbs/day}} \times 100$

   $x = 5.6\%$

10. $x = \dfrac{(7400 \text{ gpd})(8.34 \text{ lbs/gal})(6.2)(72)}{100 \quad 100}$

   $= 2755$ lbs/day VS

11. $5.7 = \dfrac{x \text{ lbs/day solids}}{35,400 \text{ lbs/day sludge}} \times 100$

   $x = 2018$ lbs/day

12. $0.5 = \dfrac{x \text{ lbs Polymer}}{(75 \text{ gal})(8.34 \text{ lbs/gal}) + x \text{ lbs}} \times 100$

   $0.5 = \dfrac{x \text{ lbs}}{625.5 \text{ lbs} + x \text{ lbs}} \times 100$

   $\dfrac{(0.5)(625.5 + x)}{100} = x$

   $3.13 + 0.005\,x = x$

   $3.13 = x - 0.005\,x$

   $3.13 = 0.995\,x$

   $3.1 \text{ lbs} = x$

CHAPTER 6 ACHIEVEMENT TEST—Cont'd

13. $25 = \dfrac{x \text{ gal}}{190,000 \text{ gal}} \times 100$

 $x = 47,500 \text{ gal}$

14. $x = (6800 \text{ lbs/day solids})(0.71)$

 $x = 4828 \text{ lbs/day VS}$

15. $x = \dfrac{(8 \text{ lbs})\left(\dfrac{9}{100}\right) + (90 \text{ lbs})\left(\dfrac{1}{100}\right)}{8 \text{ lbs} + 90 \text{ lbs}} \times 100$

 $x = \dfrac{0.7 \text{ lbs} + 0.9 \text{ lbs}}{98 \text{ lbs}} \times 100$

 $x = 1.6\%$

16. $x = \dfrac{38 \text{ hp}}{50 \text{ hp}} \times 100$

 $x = 76\% \text{ Effic.}$

17. $67 = \dfrac{x \text{ whp}}{75 \text{ mhp}} \times 100$

 $x = 50 \text{ whp}$

Chapter 7

PRACTICE PROBLEMS 7.1

1. $8.9 \text{ lbs}/1 \text{ gal} = 8.9 \text{ lbs/gal}$

2. $\dfrac{67.1 \text{ lbs/cu ft}}{7.48 \text{ gal/cu ft}} = 9.0 \text{ lbs/gal}$

3. $\dfrac{55 \text{ lbs/cu ft}}{62.4 \text{ lbs/cu ft}} = 0.9$

4. $(1.3)(8.34 \text{ lbs/gal}) = 10.8 \text{ lbs/gal}$

5. $\dfrac{8.9 \text{ lbs/gal}}{8.34 \text{ lbs/gal}} = 1.1$

PRACTICE PROBLEMS 7.2

1. $\dfrac{75 \text{ lbs}}{(30 \text{ in.})(18 \text{ in.})} = 0.1$ lbs/sq in.

2. $(2.6 \text{ lbs/sq in.})(4) = 10.4$ lbs/sq in.

3. $\dfrac{210 \text{ lbs}}{(3 \text{ ft})(3 \text{ ft})} = 23$ lbs/sq ft; $\qquad \dfrac{210 \text{ lbs}}{(3 \text{ ft})(1.25 \text{ ft})} = 56$ lbs/sq ft

4. $(6 \text{ ft})(62.4 \text{ lbs/cu ft}) = 374$ lbs/sq ft

   An alternate method: $\qquad \dfrac{6 \text{ ft}}{2.31 \text{ ft/psi}} = 2.597$ psi

   $$\dfrac{(2.597 \text{ psi}) (144 \text{ sq in.})}{\text{sq ft}} = 374 \text{ lbs/sq ft}$$

5. $\dfrac{8 \text{ ft}}{2.31 \text{ ft/psi}} = 3.5$ psi

6. First calculate the psi for water:

   $$\dfrac{4.5 \text{ ft}}{2.31 \text{ ft/psi}} = 1.9 \text{ psi}$$

   Then determine psi for a different specific gravity:

   $$(1.9 \text{ psi})(1.4) = 2.7 \text{ psi}$$

7. First determine the pressure at 10 ft depth:

   $$\dfrac{10 \text{ ft}}{2.31 \text{ ft/psi}} = 4.3 \text{ psi}$$

   Then calculate total force:

   $$(4.3 \text{ psi})(240 \text{ in.})(144 \text{ in.}) = 148,608 \text{ lbs}$$

8. First determine the pressure in lbs/sq ft at the average depth:

   $$(5)(62.4 \text{ lbs/sq ft}) = 312 \text{ lbs/sq ft}$$

   Then calculate total force on the wall:

   $$(312 \text{ lbs/sq ft})(25 \text{ ft})(10 \text{ ft}) = 78,000 \text{ lbs}$$

PRACTICE PROBLEMS 7.2—Cont'd

9. $\dfrac{(2)(9 \text{ ft})}{3} = 6 \text{ ft}$

10. First determine the lbs/sq ft pressure at the small cylinder:

$$\frac{50 \text{ lbs}}{(0.785)(0.67 \text{ ft})(0.67 \text{ ft})} = 142 \text{ lbs/sq ft}$$

Then calculate total force at the large cylinder:

$$(142 \text{ lbs/sq ft})(0.785)(2 \text{ ft})(2 \text{ ft}) = 446 \text{ lbs}$$

11. 32 psi gage + 14.7 psi atmos. = 46.7 psi absolute

PRACTICE PROBLEMS 7.3

1. 322 ft – 239 ft = 83 ft

2. Total static head is:

    790 ft – 614 ft = 176 ft

    Then TDH is:

    176 ft + 16 ft = 192 ft TDH

3. First calculate static head in psi:

    162 psi – 95 psi = 67 psi

    Then convert to ft of head:

    (67 psi)(2.31 ft/psi) = 155 ft

    And TDH is:

    155 ft + 12 ft = 167 ft TDH

4. $\dfrac{(10.11 \text{ ft head loss})(10 \text{ sections})}{100 \text{ ft section}} = 101 \text{ ft head loss}$ of 100 ft

5. There is a friction loss given for 1400 gpm and for 1500 gpm. To estimate the friction loss for 1450 gpm, use the value midway between the two values:

Friction loss for 1400 gpm—1.87 ft/100-ft section
Friction loss for 1500 gpm—2.13 ft/100-ft section

The friction loss value midway between these two values is:

$$\frac{1.87 \text{ ft} + 2.13 \text{ ft}}{2} = 2 \text{ ft}$$

Then determine the friction head loss:

$$\frac{(2 \text{ ft head loss})}{100 \text{ ft section}} (20 \text{ sections}) = 40 \text{ ft of } 100 \text{ ft}$$

6. 26 ft Equivalent Length of Pipe

PRACTICE PROBLEMS 7.4

1. $\text{Hp} = \dfrac{(40 \text{ ft})(1500 \text{ gpm})(8.34 \text{ lbs/gal})}{33,000 \text{ ft-lbs/min/hp}}$

   $= 15.2 \text{ hp}$

2. Brake horsepower is:

   $$(20 \text{ mhp})\frac{(85)}{100} = 17 \text{ bhp}$$

   Water horsepower is:

   $$(17 \text{ bhp})\frac{(80)}{100} = 13.6 \text{ whp}$$

3. $\dfrac{35 \text{ whp}}{\dfrac{85}{100}} = 41.2 \text{ bhp}$

4. $\text{Hp} = \dfrac{(60 \text{ ft})(1500 \text{ gpm})(8.34 \text{ lbs/gal})(1.2 \text{ sp. gr.})}{33,000 \text{ ft–lbs/min/hp}}$

   $= 10.9 \text{ whp}$

5. $(45 \text{ hp})(746 \text{ watts/hp}) = 33,570 \text{ watts}$

   Then convert watts to kilowatts:

   $$\frac{33,570 \text{ watts}}{1000 \text{ watts/kW}} = 33.6 \text{ kW}$$

PRACTICE PROBLEMS 7.4—Cont'd

6. First calculate kilowatts required:

   (75 mhp)(746 watts/hp) = 55,950 watts

   Then convert watts to kilowatts:

   $$\frac{55{,}950 \text{ watts}}{1000 \text{ watts/kW}} = 55.9 \text{ kw}$$

   The power consumption for the week is:

   (55.9 kw)(144 hrs) = 8049.6 kWh

   And the power cost is therefore:

   (8049.6 kWh)($0.06125/kWh) = $493.04

PRACTICE PROBLEMS 7.5

1. $$\frac{(12 \text{ ft})(10 \text{ ft})(2.6 \text{ ft})(7.48 \text{ gal/cu ft})}{5 \text{ min}} = 467 \text{ gpm}$$

2. $$\frac{55 \text{ gallons}}{0.48 \text{ min}} = 115 \text{ gpm}$$

3. The rise in level is equivalent to a gpm pumping rate of:

   $$\frac{(8 \text{ ft})(10 \text{ ft})(0.17 \text{ ft})(7.48 \text{ gal/cu ft})}{5 \text{ min}} = 20 \text{ gpm}$$

   The pumping rate is therefore:

   400 gpm – 20 gpm = 380 gpm

4. $$\frac{(0.75 \text{ gal})(30 \text{ strokes})}{\text{stroke} \quad \text{min}} = 22.5 \text{ gpm}$$

5. First calculate gpm pumping rate:

   $$\left[(0.785)(0.67 \text{ ft})(0.67 \text{ ft})(0.33 \text{ ft})(7.48 \text{ gal/cu ft})\right]\left[40 \text{ strokes/min}\right]$$
   $$= 34.8 \text{ gpm}$$

   Then convert gpm pumping rate to gpd, based on total minutes of operation:

   (34.8 gpm)(150 min/day) = 5220 gpd

CHAPTER 7 ACHIEVEMENT TEST

1. $\dfrac{66 \text{ lbs/cu ft}}{7.48 \text{ gal/cu ft}} = 8.82 \text{ lbs/gal}$

2. $\dfrac{(12 \text{ ft})(10 \text{ ft})(1.5 \text{ ft})(7.48 \text{ gal/cu ft})}{5 \text{ min}} = 269 \text{ gpm}$

3. $\dfrac{90 \text{ lbs}}{(36 \text{ in.})(24 \text{ in.})} = 0.1 \text{ lbs/sq in.}$

4. $\left[(0.785)(0.5 \text{ ft})(0.5 \text{ ft})(0.2 \text{ ft})(7.48 \text{ gal/cu ft})\right]\left[55 \dfrac{\text{strokes}}{\text{min}}\right]$

    $= 16 \text{ gpm}$

5. $\dfrac{(10 \text{ ft})(12 \text{ ft})(1.2 \text{ ft})(7.48 \text{ gal/cu ft})}{5 \text{ min}} = 215 \text{ gpm}$

6. $\dfrac{(70 \text{ ft})(900 \text{ gpm})(8.34 \text{ lb/gal})}{33{,}000 \text{ ft-lbs/min/hp}} = 15.9 \text{ hp}$

7. First calculate gpm pumping rate:

    $\left[(0.785)(0.67 \text{ ft})(0.67 \text{ ft})(0.25 \text{ ft})(7.48 \text{ gal/cu ft})\right]\left[50 \text{ strokes/min}\right]$

    $= 32.9 \text{ gpm}$

    Pumping rate is:

    $(32.9 \text{ gpm})(125 \dfrac{\text{min}}{\text{day}}) = 4113 \text{ gpd}$

8. $(1.4)(8.34 \text{ lbs/gal}) = 11.7 \text{ lbs/gal}$

9. $\dfrac{7 \text{ ft}}{2.31 \text{ ft/psi}} = 3.0 \text{ psi}$

10. Total static head is:

    $852 \text{ ft} - 760 \text{ ft} = 92 \text{ ft}$

    And TDH is:

    $92 \text{ ft} + 10 \text{ ft} = 102 \text{ ft}$

CHAPTER 7 ACHIEVEMENT TEST—Cont'd

11. 3.5 ft equivalent length of pipe

12. The pressure at the average depth is:

    (4.5 ft)(62.4 lbs/cu ft) = 281 lbs/sq ft

    The total force is:

    (281 lbs/sq ft)(15 ft)(9 ft) = 37,935 lbs

13. Brake horsepower is:

    $$(50 \text{ mhp})\frac{(90)}{100} = 45 \text{ bhp}$$

    Water horsepower is:

    $$(45 \text{ bhp})\frac{(85)}{100} = 38 \text{ whp}$$

14. $\dfrac{(2.97 \text{ ft head loss})(30 \text{ sections})}{100 \text{ ft section}}$ of 100 ft $= 89$ ft head loss

15. The friction head loss at 240 gpm is 0.87 ft/100-ft section. The head loss for the 1200-ft section is therefore:

    $\dfrac{(0.87 \text{ ft head loss})(12 \text{ sections})}{100 \text{ ft section}}$ of 100 ft $= 10.4$ ft

16. First calculate watts required:

    (60 mhp)(746 watts/hp) = 44,760 watts

    Then convert watts to kilowatts:

    $$\frac{44,760 \text{ watts}}{1000 \text{ watts/kW}} = 44.8 \text{ kW}$$

    The power used during the week is:

    (44.8 kW)(120 hrs) = 5376 kWh

    And the cost is:

    (5376 kWh)($0.05626/kWh) = $302.45

Chapter 8

PRACTICE PROBLEMS 8.1

1.   (10 ft)(10 ft)(8 ft)(7.48 gal/cu ft) = 5984 gal

2.   (12 ft)(10 ft)(6 ft)(7.48 gal/cu ft) = 5386 gal

3.   (8 ft)(8 ft)(6 ft) = 384 cu ft

4.   (10 ft)(8 ft)(x ft)(7.48 gal/cu ft) = 4787 gal

     $x = 8$ ft

5.   (8 ft)(6 ft)(2.8 ft)(7.48 gal/cu ft) = 1005 gal

PRACTICE PROBLEMS 8.2

1. $\dfrac{(8\ ft)(8\ ft)(1.6\ ft)(7.48\ gal/cu\ ft)}{5\ min} = 153$ gpm

2. $\dfrac{(10\ ft)(12\ ft)(1.1\ ft)(7.48\ gal/cu\ ft)}{3\ min} = 329$ gpm

3. $\dfrac{(8\ ft)(6\ ft)(1.6\ ft)(7.48\ gal/cu\ ft)}{3\ min} = 191$ gpm

4. $\dfrac{(8\ ft)(7\ ft)(0.7\ ft)(7.48\ gal/cu\ ft)}{1\ min} = 293$ gpm

PRACTICE PROBLEMS 8.3

1.   62 gpd÷7.48 gal/cu ft = 8.3 cu ft/day

2. $\dfrac{271\ gal/wk}{(7.48\ gal/cu\ ft)(7\ days/wk)} = 5.2$ cu ft/day

3. $\dfrac{4.7\ cu\ ft/day}{2.72\ MGD} = 1.7$ cu ft/MG

4. $\dfrac{\dfrac{78\ gal/day}{7.48\ gal/cu\ ft}}{4.6\ MGD} = 2.3$ cu ft/MG

PRACTICE PROBLEMS 8.3—Cont'd

5. $\dfrac{\dfrac{45 \text{ gal/day}}{7.48 \text{ gal/cu ft}}}{2.17 \text{ MGD}}$ = 2.8 cu ft/MG

PRACTICE PROBLEMS 8.4

1. $\dfrac{500 \text{ cu ft}}{2.8 \text{ cu ft/day}}$ = 179 days

2. $\dfrac{(8 \text{ cu yds}) (27 \text{ cu ft/cu yd})}{1.4 \text{ cu ft/day}}$ = 154 days

3. First calculate the cu ft/day screenings generated:

    (2.1 cu ft/MG) (2.7 MGD) = 5.7 cu ft/day

    Then calculate fill time:

    $\dfrac{290 \text{ cu ft}}{5.7 \text{ cu ft/day}}$ = 51 days

4. $\dfrac{x \text{ cu ft}}{3.4 \text{ cu ft/day}}$ = 120 days

    x = 408 cu ft

PRACTICE PROBLEMS 8.5

1. Since fps velocity is desired, the flow rate must be expressed as cfs

    (3 ft)(1.6 ft)(x fps) = $\dfrac{1750 \text{ gpm}}{(7.48 \text{ gal/cu ft})(60 \text{ sec/min})}$

    x = 0.8 fps

2. 25 ft/31 sec = 0.8 ft/sec

3. (4 ft)(1.1 ft)(x ft/sec) = 4.1 cfs

    x = 0.9 fps

4. 35 ft/31 sec = 1.1 ft/sec

5. Since fps velocity is desired, the flow rate must be expressed as cfs

$$(2.7 \text{ ft})(1.25 \text{ ft})(x \text{ fps}) = \frac{(1120 \text{ gpm})}{(7.48 \text{ gal/cu ft})(60 \text{ sec/min})}$$

$$x = 0.7 \text{ fps}$$

PRACTICE PROBLEMS 8.6

1. $\dfrac{11 \text{ cu ft/day}}{7 \text{ MGD}} = 1.6 \text{ cu ft/MG}$

2. $\dfrac{240 \text{ gal/day}}{(7.48 \text{ gal/cu ft})(10.3 \text{ MGD})} = 3.1 \text{ cu ft/MG}$

3. First calculate the grit removed per day:

$$\frac{(2.8 \text{ cu ft})(3.6 \text{ MGD}}{\text{MG}} = 10.1 \text{ cu ft/day}$$

Then calculate the cu ft grit removed per month:

$$\frac{(10.1 \text{ cu ft})(30 \text{ days})}{\text{day} \quad \text{month}} = 303 \text{ cu ft/month}$$

And finally, determine the cu yds removed per month:

$$\frac{303 \text{ cu ft/month}}{27 \text{ cu ft/cu yd}} = 11.2 \text{ cu yds/month}$$

4. First calculate cu ft required:

$$\frac{(2.1 \text{ cu ft})(4.12 \text{ MGD})(90 \text{ days})}{\text{MG}} = 779 \text{ cu ft required}$$

Then convert this to cu yds required:

$$\frac{779 \text{ cu ft}}{27 \text{ cu ft/cu yd}} = 29 \text{ cu yds}$$

PRACTICE PROBLEMS 8.7

1. $(2.5 \text{ ft})(1.25 \text{ ft})(0.9 \text{ fps}) = 2.8 \text{ cfs}$

2. $(2 \text{ ft})(1.3 \text{ ft})(1.2 \text{ fps})(7.48 \text{ gal/cu ft})(60 \text{ sec/min})(1440 \text{ min/day}) = 2,016,369 \text{ gpd}$

3. $(2.8 \text{ ft})(0.75 \text{ ft})(0.85 \text{ fps}) = 1.8 \text{ cfs}$

PRACTICE PROBLEMS 8.7—Cont'd

4. 0.273 cfs

5. 0.2310 MGD

6. 0.4653 cfs

7. (4.833 gps)(60 sec/min) = 290 gpm

8. The flow rate read from the nomograph is estimated as 8.6 cfs

9. The estimated flow rate from the nomograph (before using the correction factor) is 9 cfs.

   The correction factor is estimated to be 0.88:

   (9 cfs)(0.88) = 7.9 cfs

CHAPTER 8 ACHIEVEMENT TEST

1. $$\frac{260 \text{ cu ft}}{(2.4 \text{ cu ft/MG})(2.9 \text{ MGD})} = 37 \text{ days}$$

2. $$\frac{200 \text{ gal}}{(7.48 \text{ gal/cu ft})(7 \text{ days})} = 3.8 \text{ cu ft/day}$$

3. $$\frac{5.2 \text{ cu ft/day}}{2.84 \text{ MGD}} = 1.8 \text{ cu ft/MG}$$

4. $$\frac{(10 \text{ cu yds})(27 \text{ cu ft/cu yd})}{2.3 \text{ cu ft/day}} = 117 \text{ days}$$

5. 35 ft/28 sec = 1.3 ft/sec

6. 0.169 cfs

7. 0.22 MGD

8. (2.5 ft)(1.2 ft)(0.9 fps) = 2.7 cfs

9. $\dfrac{210 \text{ gallons}}{(7.48 \text{ gal/cu ft})(8.6 \text{ MGD})}$ = 3.3 cu ft/MG

10. (2.5 ft)(1.3 ft)(1.7 ft/sec)(7.48 gal/cu ft)(60 sec/min) = 2480 gpm

11. First calculate cu ft grit removal:

$\dfrac{2.2 \text{ cu ft}}{\text{MG}}$ (3.54 MGD)(30 days) = 234 cu ft

Then determine cu yds grit removal:

$\dfrac{234 \text{ cu ft}}{27 \text{ cu ft/cu yd}}$ = 8.7 cu yds

12. (3 ft)(0.9 ft)(1 fps) = 2.7 cfs

Chapter 9

PRACTICE PROBLEMS 9.1

1. $\dfrac{150,000 \text{ gal}}{71,667 \text{ gph}}$ = 2.1 hrs

2. $\dfrac{(80 \text{ ft})(20 \text{ ft})(12 \text{ ft})(7.48 \text{ gal/cu ft})}{131,250 \text{ gph}}$ = 1.1 hrs

3. $\dfrac{(0.785)(80 \text{ ft})(80 \text{ ft})(12 \text{ ft})(7.48 \text{ gal/cu ft})}{180,833 \text{ gph}}$ = 2.5 hrs

4. $\dfrac{(70 \text{ ft})(30 \text{ ft})(12 \text{ ft})(7.48 \text{ gal/cu ft})}{130,000 \text{ gph}}$ = 1.4 hrs

PRACTICE PROBLEMS 9.2

1. $\dfrac{1,463,000 \text{ gpd}}{110 \text{ ft}} = 13,300 \text{ gpd/ft}$

2. $\dfrac{2,950,000 \text{ gpd}}{(3.14)(70 \text{ ft})} = 13,421 \text{ gpd/ft}$

3. $\dfrac{(2440 \text{ gpm})(1440 \text{ min/day})}{(3.14)(80 \text{ ft})} = 13,987 \text{ gpd/ft}$

4. $\dfrac{1,780,000 \text{ gpd}}{188 \text{ ft}} = 9468 \text{ gpd/ft}$

PRACTICE PROBLEMS 9.3

1. $\dfrac{2,890,000 \text{ gpd}}{(0.785)(60 \text{ ft})(60 \text{ ft})} = 1023 \text{ gpd/sq ft}$

2. $\dfrac{2,160,000 \text{ gpd}}{(75 \text{ ft})(25 \text{ ft})} = 1152 \text{ gpd/sq ft}$

3. $\dfrac{2,590,000 \text{ gpd}}{(90 \text{ ft})(40 \text{ ft})} = 719 \text{ gpd/sq ft}$

4. $\dfrac{(2580 \text{ gpm})(1440 \text{ min/day})}{(0.785)(70 \text{ ft})(70 \text{ ft})} = 966 \text{ gpd/sq ft}$

PRACTICE PROBLEMS 9.4

1. $\dfrac{(2940 \text{ mg/}L)(3.9 \text{ MGD})(8.34 \text{ lbs/gal})}{(0.785)(75 \text{ ft})(75 \text{ ft})} = 21.7 \text{ lbs/day/sq ft}$

2. $\dfrac{(3140 \text{ mg/}L)(4 \text{ MGD})(8.34 \text{ lbs/gal})}{(0.785)(75 \text{ ft})(75 \text{ ft})} = 23.7 \text{ lbs/day/sq ft}$

3. $\dfrac{(2610 \text{ mg/}L)(3.13 \text{ MGD})(8.34 \text{ lbs/gal})}{(0.785)(60 \text{ ft})(60 \text{ ft})} = 24.1 \text{ lbs/day/sq ft}$

4. $\dfrac{(3260 \text{ mg/}L)(2.87 \text{ MGD})(8.34 \text{ lbs/gal})}{(0.785)(70 \text{ ft})(70 \text{ ft})} = 20.3 \text{ lbs/day/sq ft}$

PRACTICE PROBLEMS 9.5

1. (125 mg/*L*)(5.16 MGD)(8.34 lbs/gal) = 5379 lbs/day SS

2. (45 mg/*L* Rem.)(2.84 MGD)(8.34 lbs/gal) = 1066 lbs/day SS

3. (87 mg/*L* Rem.)(4.17 MGD)(8.34 lbs/gal) = 3026 lbs/day BOD

4. (186 mg/*L* Rem.)(0.96 MGD)(8.34 lbs/gal) = 1489 lbs/day SS

PRACTICE PROBLEMS 9.6

1. $\dfrac{123 \text{ mg/}L \text{ Rem.}}{220 \text{ mg/}L}$ x 100 = 56% Removal Efficiency

2. $\dfrac{117 \text{ mg/}L \text{ Rem.}}{195 \text{ mg/}L}$ x 100 = 60% Removal Efficiency

3. $\dfrac{211 \text{ mg/}L \text{ Rem.}}{270 \text{ mg/}L}$ x 100 = 78% Removal Efficiency

4. $\dfrac{118 \text{ mg/}L \text{ Rem.}}{310 \text{ mg/}L}$ x 100 = 38% Removal Efficiency

CHAPTER 9 ACHIEVEMENT TEST

1. $\dfrac{(0.785)(75 \text{ ft})(75 \text{ ft})(12 \text{ ft})(7.48 \text{ gal/cu ft})}{164,167 \text{ gph}}$ = 2.4 hrs

2. $\dfrac{2,260,000 \text{ gpd}}{(0.785)(50 \text{ ft})(50 \text{ ft})}$ = 1152 gpd/sq ft

3. $\dfrac{3,728,000 \text{ gpd}}{210 \text{ ft}}$ = 17,752 gpd/sq ft

4. $\dfrac{(2640 \text{ mg/}L)(2.415 \text{ MGD})(8.34 \text{ lbs/gal})}{(0.785)(55 \text{ ft})(55 \text{ ft})}$ = 22.4 lbs/day/sq ft

5. $\dfrac{2,620,000 \text{ gpd}}{(0.785)(60 \text{ ft})(60 \text{ ft})}$ = 927 gpd/sq ft

6. $\dfrac{(2815 \text{ mg/}L)(3.53 \text{ MGD})(8.34 \text{ lbs/gal})}{(0.785)(70 \text{ ft})(70 \text{ ft})}$ = 21.5 lbs/day/sq ft

CHAPTER 9 ACHIEVEMENT TEST

7. (113 mg/*L* Rem.)(5.1 MGD)(8.34 lbs/gal) = 4806 lbs/day Rem.

8. $\dfrac{(80 \text{ ft})(30 \text{ ft})(14 \text{ ft})(7.48 \text{ gal/cu ft})}{168,750 \text{ gph}}$ = 1.5 hrs

9. $\dfrac{(1920 \text{ gpm})(1440 \text{ min/day})}{(3.14)(60 \text{ ft})}$ = 14,675 gpd/ft

10. (81 mg/*L* Rem.)(4.36 MGD)(8.34 lbs/gal) = 2945 lbs/day BOD removed

11. (149 mg/*L* Rem.)(3.76 MGD)(8.34 lbs/gal) = 4672 lbs/day Removed

12. $\dfrac{182 \text{ mg/}L \text{ Removed}}{250 \text{ mg/}L}$ x 100 = 73% Removal Efficiency

13. $\dfrac{2,125,000 \text{ gpd}}{(80 \text{ ft})(35 \text{ ft})}$ = 759 gpd/sq ft

Chapter 10

PRACTICE PROBLEMS 10.1

1. $\dfrac{750,000 \text{ gpd}}{(0.785)(75 \text{ ft})(75 \text{ ft})}$ = 170 gpd/sq ft

2. $\dfrac{(2310 \text{ gpm})(1440 \text{ min/day})}{(0.785)(85 \text{ ft})(85 \text{ ft})}$ = 586 gpd/sq ft

3. $\dfrac{1,400,000 \text{ gpd}}{(0.785)(90 \text{ ft})(90 \text{ ft})}$ = 220 gpd/sq ft

4. First calculate the acres surface area:

$\dfrac{(0.785)(95 \text{ ft})(95 \text{ ft})}{43,560 \text{ sq ft/ac}}$ = 0.16 ac

Then calculate hydraulic loading:

$\dfrac{1.9 \text{ MGD}}{0.16 \text{ ac}}$ = 11.9 MGD/ac

PRACTICE PROBLEMS 10.2

1. $\dfrac{(205 \text{ mg}/L)(1.2 \text{ MGD})(8.34 \text{ lbs/gal})}{31.8 \quad 1000\text{-cu ft}} = 64.5$ lbs BOD/day/1000 cu ft

2. $\dfrac{(110 \text{ mg}/L)(3.24 \text{ MGD})(8.34 \text{ lbs/gal})}{30.1 \quad 1000\text{-cu ft}} = 98.8$ lbs BOD/day/1000 cu ft

3. $\dfrac{(195 \text{ mg}/L)(0.8 \text{ MGD})(8.34 \text{ lbs/gal})}{23.1 \quad 1000\text{-cu ft}} = 56.3$ lbs BOD/day/1000 cu ft

4. First calculate ac-ft media:

$$\dfrac{(0.785)(95 \text{ ft})(95 \text{ ft})(6 \text{ ft})}{43,560 \text{ cu ft/ac-ft}} = 1.0 \text{ ac-ft}$$

Then determine organic loading:

$$\dfrac{(130 \text{ mg}/L)(1.3 \text{ MGD})(8.34 \text{ lbs/gal})}{1 \text{ ac-ft}} = \dfrac{1409 \text{ lbs BOD/day}}{\text{ac-ft}}$$

PRACTICE PROBLEMS 10.3

1. (121 mg/L Rem.)(3.178 MGD)(8.34 lbs/gal) = 3207 lbs/day SS Rem.

2. (168 mg/L Rem.)(1.61 MGD)(8.34 lbs/gal) = 2256 lbs/day BOD Rem.

3. (177 mg/L BOD Rem.)(2.84 MGD)(8.34 lbs/gal) = 4192 lbs/day BOD Rem.

4. (189 mg/L BOD Rem.)(5.1 MGD)(8.34 lbs/gal) = 8039 lbs/day BOD Rem.

PRACTICE PROBLEMS 10.4

1. $\dfrac{100 \text{ mg}/L \text{ Removed}}{146 \text{ mg}/L} \times 100 = 68\%$ Removal Efficiency

2. $\dfrac{232 \text{ mg}/L \text{ Removed}}{252 \text{ mg}/L} \times 100 = 92\%$ Removal Efficiency

3. $\dfrac{173 \text{ mg}/L \text{ Removed}}{197 \text{ mg}/L} \times 100 = 88\%$ Removal Efficiency

4. $\dfrac{86 \text{ mg}/L \text{ Removed}}{110 \text{ mg}/L} \times 100 = 78\%$ Removal Efficiency

PRACTICE PROBLEMS 10.5

1. $\dfrac{3.4 \text{ MGD}}{3.2 \text{ MGD}} = 1.1$

2. $\dfrac{2.12 \text{ MGD}}{1.52 \text{ MGD}} = 1.4$

3. $\dfrac{3.73 \text{ MGD}}{2.62 \text{ MGD}} = 1.4$

4. $1.4 = \dfrac{x \text{ MGD}}{4.4 \text{ MGD}}$

    $x = 6.2 \text{ MGD}$

CHAPTER 10 ACHIEVEMENT TEST

1. $\dfrac{660,000 \text{ gpd}}{(0.785)(90 \text{ ft})(90 \text{ ft})} = 104 \text{ gpd/sq ft}$

2. $\dfrac{(160 \text{ mg/}L)(1.2 \text{ MGD})(8.34 \text{ lbs/gal})}{28.4 \quad 1000\text{-cu ft}} = 56.4 \text{ lbs BOD/day/1000 cu ft}$

3. $(113 \text{ mg/}L \text{ Rem.})(2.668 \text{ MGD})(8.34 \text{ lbs/gal}) = 2514 \text{ lbs/day SS Rem.}$

4. $\dfrac{143 \text{ mg/}L \text{ Rem.}}{210 \text{ mg/}L} \times 100 = 68\% \text{ Removal Efficiency}$

5. $(148 \text{ mg/}L \text{ Rem.})(1.33 \text{ MGD})(8.34 \text{ lbs/gal}) = 1642 \text{ lbs/day Rem.}$

6. $\dfrac{2,750,000 \text{ gpd}}{(0.785)(80 \text{ ft})(80 \text{ ft})} = 547 \text{ gpd/sq ft}$

7. $\dfrac{185 \text{ mg/}L \text{ Rem.}}{205 \text{ mg/}L} \times 100 = 90\% \text{ Removal Efficiency}$

8. $\dfrac{(139 \text{ mg/}L)(2.18 \text{ MGD})(8.34 \text{ lbs/gal})}{35.2 \quad 1000\text{-cu ft}} = 72 \text{ lbs BOD/day/1000 cu ft}$

9. First calculate ac-ft media:

$$\frac{(0.785)(90 \text{ ft})(90 \text{ ft})(7 \text{ ft})}{43,560 \text{ cu ft/ac-ft}} = 1.0 \text{ ac-ft}$$

Then determine loading rate:

$$\frac{(140 \text{ mg/}L)(1.05 \text{ MGD})(8.34 \text{ lbs/gal})}{1 \text{ ac-ft}} = 1226 \text{ lbs BOD/day/ac-ft}$$

10. $(178 \text{ mg/}L \text{ Rem.})(4.11 \text{ MGD})(8.34 \text{ lbs/gal}) = 6101 \text{ lbs/day Rem.}$

11. $\dfrac{3.6 \text{ MGD}}{3.3 \text{ MGD}} = 1.1$

12. First determine acres surface area:

$$\frac{(0.785)(90 \text{ ft})(90 \text{ ft})}{43,560 \text{ sq ft/ac}} = 0.15 \text{ ac}$$

Then calculate loading:

$$\frac{1.7 \text{ MGD}}{0.15 \text{ ac}} = 11.3 \text{ MGD/ac}$$

13. $\dfrac{1,890,000 \text{ gpd}}{(0.785)(80 \text{ ft})(80 \text{ ft})} = 376 \text{ gpd/sq ft}$

14. $\dfrac{(167 \text{ mg/}L)(0.78 \text{ MGD})(8.34 \text{ lbs/gal})}{23.1 \qquad 1000\text{-cu ft}} = 47 \text{ lbs BOD/day/1000 cu ft}$

15. $\dfrac{2.27 \text{ MGD}}{1.61 \text{ MGD}} = 1.4$

16. $\dfrac{203 \text{ mg/}L \text{ Rem.}}{236 \text{ mg/}L} \times 100 = 86\% \text{ Removal Efficiency}$

Chapter 11

PRACTICE PROBLEMS 11.1

1. $\dfrac{2,940,000 \text{ gpd}}{700,000 \text{ sq ft}} = 4.2 \text{ gpd/sq ft}$

2. $\dfrac{4,654,000 \text{ gpd}}{850,000 \text{ sq ft}} = 5.5 \text{ gpd/sq ft}$

3. $\dfrac{1,490,000 \text{ gpd}}{400,000 \text{ sq ft}} = 3.7 \text{ gpd/sq ft}$

4. $\dfrac{x \text{ gpd}}{750,000 \text{ sq ft}} = 6 \text{ gpd/sq ft}$

   $x = 4,500,000 \text{ gpd}$

PRACTICE PROBLEMS 11.2

1. (238 mg/*L* SS)(0.45) = 107 mg/*L* Particulate BOD
                        K-value

2. 218 mg/*L* Tot. BOD = (232 mg/*L*)(0.5) + *x* mg/*L* Sol. BOD
                        Partic. BOD
   x = 102 mg/*L* Sol. BOD

3. 235 mg/*L* Tot. BOD = (140 mg/*L*)(0.5) + *x* mg/*L* Sol. BOD
                        Partic. BOD
   x = 165 mg/*L* Sol. BOD

4. First calculate mg/*L* Sol. BOD:

   275 mg/*L* Tot. BOD = (264 mg/*L*)(0.6) + *x* mg/*L* Sol. BOD
                        Particulate BOD

   x = 117 mg/*L* Sol. BOD

   Then calculate lbs/day Sol. BOD:

   (117 mg/*L*)(1.8 MGD)(8.34 lbs/gal) = 1756 lbs/day Sol. BOD
   Soluble BOD

PRACTICE PROBLEMS 11.3

1. $\dfrac{(155 \text{ mg}/L)(4.27 \text{ MGD})(8.34 \text{ lbs/gal})}{\dfrac{900}{} \quad \dfrac{}{1000\text{-sq ft}}} = 6.1$ lbs/day/1000 sq ft

2. $\dfrac{(174 \text{ mg}/L)(1.45 \text{ MGD})(8.34 \text{ lbs/gal})}{\dfrac{600}{} \quad \dfrac{}{1000\text{-sq ft}}} = 3.5$ lbs/day/1000 sq ft

3. $\dfrac{(124 \text{ mg}/L)(2.64 \text{ MGD})(8.34 \text{ lbs/gal})}{\dfrac{650}{} \quad \dfrac{}{1000\text{-sq ft}}} = 4.2$ lbs/day/1000 sq ft

4. First calculate soluble BOD:

    182 mg/L Total BOD = (135 mg/L)(0.45) + x mg/L Soluble BOD

    $x = 121$ mg/L Sol. BOD

    Then determine organic loading rate:

    $\dfrac{(121 \text{ mg}/L)(2.6 \text{ MGD})(8.34 \text{ lbs/gal})}{\dfrac{725}{} \quad \dfrac{}{1000\text{-sq ft}}} = 3.6$ lbs/day/1000 sq ft

CHAPTER 11 ACHIEVEMENT TEST

1. $\dfrac{2{,}820{,}000 \text{ gpd}}{650{,}000 \text{ sq ft}} = 4.3$ gpd/sq ft

2. (216 mg/L SS)(0.5) = 108 mg/L

3. $\dfrac{(149 \text{ mg}/L)(1.86 \text{ MGD})(8.34 \text{ lbs/gal})}{\dfrac{700}{} \quad \dfrac{}{1000\text{-sq ft}}} = 3.3$ lbs/day/1000 sq ft

4. First determine mg/L soluble BOD:

    195 mg/L Total BOD = (205 mg/L)(0.6) + x mg/L Sol. BOD

    $x = 72$ mg/L Sol. BOD

    Then calculate lbs/day soluble BOD:

    (72 mg/L)(2.7 MGD)(8.34 lbs/gal) = 1621 lbs/day

CHAPTER 11 ACHIEVEMENT TEST—Cont'd

5. $\dfrac{4{,}276{,}000 \text{ gpd}}{900{,}000 \text{ sq ft}} = 4.8 \text{ gpd/sq ft}$

6. $\dfrac{(119 \text{ mg/}L)(2.316 \text{ MGD})(8.34 \text{ lbs/gal})}{720 \qquad 1000\text{-sq ft}} = 3.2 \text{ lbs/day/1000 sq ft}$

Chapter 12

PRACTICE PROBLEMS 12.1

1. (70 ft)(30 ft)(15 ft)(7.48 gal/cu ft) = 235,620 gal

2. (90 ft)(35 ft)(14 ft)(7.48 gal/cu ft) = 329,868 gal

3. (0.785)(70 ft)(70 ft)(12 ft)(7.48 gal/cu ft) = 345,262 gal

4. (0.785)(60 ft)(60 ft)(14 ft)(7.48 gal/cu ft) = 295,939 gal

5. $(6.5 \text{ ft})(4 \text{ ft})\big[480 \text{ ft} + (3.14)(90 \text{ ft})\big]$

   $= (6.5 \text{ ft})(4 \text{ ft})(762.6 \text{ ft})$

   $= 19{,}828 \text{ cu ft}$

6. $(5.5 \text{ ft})(5 \text{ ft})\big[380 \text{ ft} + (3.14)(80 \text{ ft})\big]$

   $= (5.5 \text{ ft})(5 \text{ ft})(631.2 \text{ ft})$

   $= 17{,}358 \text{ cu ft}$

PRACTICE PROBLEMS 12.2

1. (230 mg/L)(0.89 MGD)(8.34 lbs/gal) = 1707 lbs/day BOD

2. (150 mg/L)(4.23 MGD)(8.34 lbs/gal) = 5292 lbs/day COD

3. (155 mg/L)(3.12 MGD)(8.34 lbs/gal) = 4033 lbs/day BOD

4. (140 mg/L)(4.72 MGD)(8.34 lbs/gal) = 5511 lbs/day COD

PRACTICE PROBLEMS 12.3

1. (2040 mg/L)(0.45 MG)(8.34 lbs/gal) = 7656 lbs MLSS

2. (2160 mg/*L*)(0.18 MG)(8.34 lbs/gal) = 3243 lbs MLVSS

3. (2340 mg/*L*)(0.4 MG)(8.34 lbs/gal) = 7806 lbs MLVSS

4. (2710 mg/*L*)(0.54 MG)(8.34 lbs/gal) = 12,205 lbs MLVSS

5. (2460 mg/*L*)(0.47 MG)(8.34 lbs/gal)(0.72) = 6,943 lbs MLVSS

PRACTICE PROBLEMS 12.4

1. $\dfrac{(195 \text{ mg/}L)(2.61 \text{ MGD})(8.34 \text{ lbs/gal})}{(2560 \text{ mg/}L)(0.47 \text{ MG})(8.34 \text{ lbs/gal})} = 0.42$

2. $\dfrac{(145 \text{ mg/}L)(3.26 \text{ MGD})(8.34 \text{ lbs/gal})}{(2490 \text{ mg/}L)(0.47 \text{ MG})(8.34 \text{ lbs/gal})} = 0.40$

3. $\dfrac{(175 \text{ mg/}L)(0.31 \text{ MGD})(8.34 \text{ lbs/gal})}{(2576 \text{ mg/}L)(0.19 \text{ MG})(8.34 \text{ lbs/gal})} = 0.11$

4. $\dfrac{(178 \text{ mg/}L)(3.1 \text{ MGD})(8.34 \text{ lbs/gal})}{x \text{ lbs MLVSS}} = 0.7$

    x = 6574 mg/*L* MLVSS

5. $\dfrac{(139 \text{ mg/}L)(2.46 \text{ MGD})(8.34 \text{ lbs/gal})}{x \text{ lbs MLVSS}} = 0.3$

    x = 9506 lbs MLVSS

PRACTICE PROBLEMS 12.5

1. $\dfrac{15,600 \text{ lbs MLSS}}{2520 \text{ lbs/day SS}} = 6.2 \text{ days}$

2. $\dfrac{(2740 \text{ mg/}L)(0.47 \text{ MG})(8.34 \text{ lbs/gal})}{(108 \text{ mg/}L)(2.8 \text{ MGD})(8.34 \text{ lbs/gal})} = 4.3 \text{ days}$

3. $\dfrac{(2480 \text{ mg/}L)(0.45 \text{ MG})(8.34 \text{ lbs/gal})}{(110 \text{ mg/}L)(2.72 \text{ MGD})(8.34 \text{ lbs/gal})} = 3.7 \text{ days}$

4. $\dfrac{(2940 \text{ mg/}L)(0.50 \text{ MG})(8.34 \text{ lbs/gal})}{(105 \text{ mg/}L)(1.94 \text{ MGD})(8.34 \text{ lbs/gal})} = 7.2 \text{ days}$

5. $\dfrac{(3720 \text{ mg/}L)(0.21 \text{ MG})(8.34 \text{ lbs/gal})}{(201 \text{ mg/}L)(0.26 \text{ MGD})(8.34 \text{ lbs/gal})} = 14.9 \text{ days}$

PRACTICE PROBLEMS 12.6

1. $\dfrac{28{,}600 \text{ lbs}}{2860 \text{ lbs/day} + 390 \text{ lbs/day}} = 8.8 \text{ days}$

2. $\dfrac{(2650 \text{ mg/}L)(1.4 \text{ MG})(8.34 \text{ lbs/gal}) + (1920 \text{ mg/}L)(0.105 \text{ MG})(8.34 \text{ lbs/gal})}{(5960 \text{ mg/}L)(0.07 \text{ MGD})(8.34 \text{ lbs/gal}) + (20 \text{ mg/}L)(3.1 \text{ MGD})(8.34 \text{ lbs/gal})}$

$\dfrac{30{,}941 \text{ lbs} + 1681 \text{ lbs}}{3497 \text{ lbs/day} + 517 \text{ lbs/day}} = 8.2 \text{ days}$

3. $\dfrac{(2115 \text{ mg/}L)(0.625 \text{ MG})(8.34 \text{ lbs/gal})}{1570 \text{ lbs/day} + 230 \text{ lbs/day}} = 6.1 \text{ days}$

4. $\dfrac{(2940 \text{ mg/}L)(0.46 \text{ MG})(8.34 \text{ lbs/gal})}{(6110 \text{ mg/}L)(0.025 \text{ MGD})(8.34 \text{ lbs/gal}) + (17 \text{ mg/}L)(1.2 \text{ MGD})(8.34 \text{ lbs/gal})}$

$\dfrac{11{,}279 \text{ lbs}}{1274 \text{ lbs/day} + 170 \text{ lbs/day}} = 7.8 \text{ days}$

5. $\dfrac{(x \text{ mg/}L \text{ MLSS})(0.47 \text{ MG})(8.34 \text{ lbs/gal})}{(5740 \text{ mg/}L)(0.0284 \text{ MGD})(8.34 \text{ lbs/gal}) + (18 \text{ mg/}L)(1.76 \text{ MGD})(8.34 \text{ lbs/gal})} = 7.5 \text{ days}$

$\dfrac{(x \text{ mg/}L \text{ MLSS})(0.47 \text{ MG})(8.34 \text{ lbs/gal})}{1360 \text{ lbs/day} + 264 \text{ lbs/day}} = 7.5 \text{ days}$

$\dfrac{(x \text{ mg/}L)(0.47 \text{ MG})(8.34 \text{ lbs/gal})}{1624 \text{ lbs/day}} = 7.5 \text{ days}$

$x = 3107 \text{ mg/}L$

6. $\dfrac{(2460 \text{ mg/}L)(1.5 \text{ MG})(8.34 \text{ lbs/gal}) + (1850 \text{ mg/}L)(0.11 \text{ MG})(8.34 \text{ lbs/gal})}{(8040 \text{ mg/}L)(0.06 \text{ MGD})(8.34 \text{ lbs/gal}) + (18 \text{ mg/}L)(3.4 \text{ MGD})(8.34 \text{ lbs/gal})}$

$\dfrac{30{,}775 \text{ lbs MLSS} + 1697 \text{ lbs CSS}}{4023 \text{ lbs/day SS} + 510 \text{ lbs/day}}$

$= 7.2 \text{ days}$

PRACTICE PROBLEMS 12.7

1. $\dfrac{215 \text{ m}L/L}{1000 \text{ m}L/L - 215 \text{ m}L/L} = 0.27$

2. Suspended Solids In, lbs/day = Suspended solids Out, lbs/day

$$\underset{\substack{mg/L}}{(2460)}(3.4 \text{ MGD} + x \text{ MGD})\underset{\substack{lbs/gal}}{(8.34)} = \underset{\substack{mg/L}}{(7850)}(x \text{ MGD})\underset{\substack{lbs/gal}}{(8.34)} + \underset{\substack{mg/L}}{(7850)}(0.06 \text{ MGD})\underset{\substack{lbs/gal}}{(8.34)}$$

The 8.34 lbs/gal factor may be divided out from each term of the equation, leaving:

$$(2460 \text{ mg/L})(3.4 \text{ MGD} + x \text{ MGD}) = (7850 \text{ mg/L})(x \text{ MGD}) + (7850 \text{ mg/L})(0.06 \text{ MGD})$$

$$8364 + 2460x = 7850x + 471$$

$$7893 = 5390x$$

$$1.46 \text{ MGD} = x$$

3. a) $\dfrac{270 \text{ mL/L}}{1000 \text{ mL/L} - 270 \text{ mL/L}} = 0.37$

   b) $\dfrac{270 \text{ RAS}}{730 \text{ Sec. Infl.}} = \dfrac{x \text{ MGD RAS}}{3.12 \text{ MGD Sec. Infl. Flow}}$

   $x = 1.15$ MGD RAS Flow

4. a) $\dfrac{285 \text{ mL/L}}{1000 \text{ mL/L} - 285 \text{ mL/L}} = 0.40$

   b) $\dfrac{285 \text{ RAS}}{715 \text{ Sec. Inf}} = \dfrac{x \text{ MGD}}{1.67 \text{ MGD}}$

   $x = 0.67$ MGD

5. Suspended Solids In, lbs/day = Suspended solids Out, lbs/day

$$(7490 \text{ mg/L})(x \text{ MGD RAS})(8.34 \text{ lbs/gal}) = (2100 \text{ mg/L})(6.3 \text{ MGD} + x \text{ MGD RAS})(8.34 \text{ lbs/gal})$$

The 8.34 lbs/gal factor may be divided out on both sides of the equation, leaving:

$$(7490)(x \text{ MGD}) = (2100)(6.3 + x \text{ MGD})$$

$$7490x = 13{,}230 + 2100x$$

$$5390x = 13{,}230$$

$$x = 2.45 \text{ MGD}$$

PRACTICE PROBLEMS 12.8

1. Calculate desired lbs MLSS, using the desired F/M ratio:

$$\dfrac{3300 \text{ lbs/day BOD}}{\dfrac{(x \text{ lbs/day MLSS})(68)}{100}} = 0.6$$

$$x = 8088 \text{ lbs MLSS desired}$$

PRACTICE PROBLEMS 12.8—Cont'd

2. First calculate the actual lbs MLSS in the aeration tank:

$$(2670 \text{ mg/}L)(0.78 \text{ MG})(8.34 \text{ lbs/gal}) = 17,369 \text{ lbs MLSS}$$

Then compare actual with desired lbs MLSS to determine wasting needed:

$$\underset{\text{actual}}{17,369 \text{ lbs MLSS}} - \underset{\text{desired}}{14,850 \text{ lbs MLSS}} = \underset{\text{to be wasted}}{2519 \text{ lbs MLSS}}$$

3. First calculate desired lbs MLSS using the desired F/M ratio:

$$\frac{(108 \text{ mg/}L)(2.93 \text{ MGD})(8.34 \text{ lbs/gal})}{(x \text{ lbs MLSS})\dfrac{(67)}{100}} = 0.3$$

$$x = 13,130 \text{ lbs MLSS desired}$$

Then calculate actual lbs MLSS:

$$(2100 \text{ mg/}L)(1.1 \text{ MG})(8.34 \text{ lbs/gal}) = 19,265 \text{ lbs MLSS actual}$$

Now lbs to be wasted can be determined:

$$19,265 \text{ lbs MLSS actual} - 13,130 \text{ lbs MLSS desired} = \underset{\text{to be wasted}}{6135 \text{ lbs MLSS}}$$

4. Calculate the desired lbs MLSS, using the desired sludge age:

$$\frac{x \text{ lbs MLSS}}{3140 \text{ lbs/day SS}} = \underset{\text{Sludge Age}}{5.5 \text{ days desired}}$$

$$x = 17,270 \text{ lbs MLSS desired}$$

Then calculate actual lbs MLSS:

$$(2920 \text{ mg/}L)(0.85 \text{ MG})(8.34 \text{ lbs/gal}) = 20,700 \text{ lbs MLSS actual}$$

Now determine the lbs MLSS determined:

$$20,700 \text{ lbs MLSS actual} - 17,270 \text{ lbs MLSS desired} = \underset{\text{to be wasted}}{3430 \text{ lbs MLSS}}$$

5.
$$\frac{32,100 \text{ lbs MLSS}}{x \text{ lbs/day WAS} + (22 \text{ mg/}L)(3.16 \text{ MGD})(8.34 \text{ lbs/gal})} = 8.5 \text{ days}$$

First simplify the equation:

$$\frac{32,100 \text{ lbs}}{x \text{ lbs/day} + 580 \text{ lbs/day}} = 8.5 \text{ days}$$

$$\frac{32,100}{8.5} = x + 580$$

$$3776 = x + 580$$

$$\underset{\text{WAS}}{3196 \text{ lbs/day}} = x$$

6. $$\frac{(2940\ mg/L)(1.33\ MG)(8.34\ lbs/gal)}{x\ lbs/day\ WAS + (15\ mg/L)(6.85\ MG)(8.34\ lbs/gal)} = 10\ days$$

$$\frac{32,611\ lbs}{x\ lbs/day + 857\ lbs/day} = 10\ days$$

$$\frac{32,611}{10} = x + 857$$

$$2404\ lbs/day = x$$
$$WAS$$

PRACTICE PROBLEMS 12.9

1. $(6630\ mg/L)(x\ MGD)(8.34\ lbs/gal) = 5640\ lbs/day\ dry\ solids$

   $x = 0.10\ MGD$

2. a) $(6120\ mg/L)(x\ MGD)(8.34\ lbs/gal) = 8640\ lbs/day\ dry\ solids$

   $x = 0.17\ MGD$

   b) $0.17\ MGD = 170,000\ gpd$

   Convert this flow rate to gpm:

   $$\frac{170,000\ gpd}{1440\ min/day} = 118\ gpm$$

3. $$\frac{(2630\ mg/L)(1.7\ MG)(8.34\ lbs/gal)}{(7410\ mg/L)(x\ MGD)(8.34\ lbs/gal)+(19\ mg/L)(4.1\ MGD)(8.34\ lbs/gal)} = 9.5\ days$$

   $$\frac{37,288\ lbs\ MLSS}{(7410\ mg/L)(x\ MGD)(8.34\ lbs/gal) + 650\ lbs/day} = 9.5\ days$$

   $$\frac{37,288}{9.5} = (7410)(x)(8.34) + 650$$

   $$3925 = (7410)(x)(8.34) + 650$$

   $$3275 = (7410)(x)(8.34)$$

   $$0.053\ MGD = x$$

PRACTICE PROBLEMS 12.9—Cont'd

4. $$\frac{(2580 \text{ mg/}L)(1.6 \text{ MG})(8.34 \text{ lbs/gal})}{(5990 \text{ mg/}L)(x \text{ MGD})(8.34 \text{ lbs/gal})+(13 \text{ mg/}L)(3.9 \text{ MGD})(8.34 \text{ lbs/gal})} = 8 \text{ days}$$

$$\frac{34,428 \text{ lbs MLSS}}{(5990 \text{ mg/}L)(x \text{ MGD})(8.34 \text{ lbs/gal}) + 423 \text{ lbs/day}} = 8 \text{ days}$$

$$\frac{34,428}{8} = (5990)(x)(8.34) + 423$$

$$4304 = (5990)(x)(8.34) + 423$$

$$3881 = (5990)(x)(8.34)$$

$$0.078 \text{ MGD} = x$$

PRACTICE PROBLEMS 12.10

1. $\dfrac{165,000 \text{ gal}}{7500 \text{ gph}} = 22 \text{ hrs}$

2. $\dfrac{360,000 \text{ gal}}{8750 \text{ gph}} = 41 \text{ hrs}$

3. $\dfrac{415,000 \text{ gal}}{12,292 \text{ gph}} = 34 \text{ hrs}$

4. $\dfrac{190,000 \text{ gal}}{12,708 \text{ gph}} = 15 \text{ hrs}$

CHAPTER 12 ACHIEVEMENT TEST

1. $(90 \text{ ft})(30 \text{ ft})(16 \text{ ft})(7.48 \text{ gal/cu ft}) = 323,136 \text{ gal}$

2. $(217 \text{ mg/}L)(1.668 \text{ MGD})(8.34 \text{ lbs/gal}) = 3019 \text{ lbs/day}$

3. $\dfrac{(218 \text{ mg/}L)(0.389 \text{ MGD})(8.34 \text{ lbs/gal})}{(3250 \text{ mg/}L)(0.21 \text{ MG})(8.34 \text{ lbs/gal})\dfrac{(67)}{100}} = 0.19$

4. $(0.785)(80 \text{ ft})(80 \text{ ft})(10 \text{ ft})(7.48 \text{ gal/cu ft}) = 375,795 \text{ gal}$

5. $\dfrac{(175 \text{ mg/}L)(2.13 \text{ MGD})(8.34 \text{ lbs/gal})}{(2880 \text{ mg/}L)(0.42 \text{ MG})(8.34 \text{ lbs/gal})} = 0.31$

6. $(6.5 \text{ ft})(3.5 \text{ ft})\big[(640 \text{ ft} + (3.14)(100 \text{ ft})\big]$

    $= (6.5 \text{ ft})(3.5 \text{ ft})(954 \text{ ft})$

    $= 21,704 \text{ cu ft}$

7. $(155 \text{ mg/}L)(3.84 \text{ MGD})(8.34 \text{ lbs/gal}) = 4964 \text{ lbs/day}$

8. $\dfrac{(2610 \text{ mg/}L)(0.525 \text{ MG})(8.34 \text{ lbs/gal})}{(185 \text{ mg/}L)(1.7 \text{ MG})(8.34 \text{ lbs/gal})} = 4.4 \text{ days}$

9. $\dfrac{(143 \text{ mg/}L)(2.78 \text{ MGD})(8.34 \text{ lbs/gal})}{x \text{ lbs MLVSS}} = 0.4$

    $x = 8289 \text{ lbs MLVSS}$

10. $\dfrac{390,000 \text{ gal}}{11,667 \text{ gph}} = 33 \text{ hrs}$

11. $\dfrac{(158 \text{ mg/}L)(2.39 \text{ MGD})(8.34 \text{ lbs/gal})}{x \text{ lbs MLVSS}} = 0.7$

    $x = 4499 \text{ lbs MLVSS}$

12. $\dfrac{(2830 \text{ mg/}L)(0.44 \text{ MGD})(8.34 \text{ lbs/gal})}{(160 \text{ mg/}L)(1.1 \text{ MG})(8.34 \text{ lbs/gal})} = 7.1 \text{ days}$

13. $\dfrac{600,000 \text{ gal}}{14,167 \text{ gph}} = 42.4 \text{ hrs}$

14. $\dfrac{(3910 \text{ mg/}L)(0.25 \text{ MGD})(8.34 \text{ lbs/gal})}{(195 \text{ mg/}L)(0.3 \text{ MGD})(8.34 \text{ lbs/gal})} = 16.7 \text{ days}$

15. $(2660 \text{ mg/}L)(0.425 \text{ MG})(8.34 \text{ lbs/gal}) = 9428 \text{ lbs SS}$

16. $\dfrac{(144 \text{ mg/}L)(2.81 \text{ MGD})(8.34 \text{ lbs/gal})}{x \text{ lbs MLVSS}} = 0.3$

    $x = 11,249 \text{ lbs MLVSS}$

CHAPTER 12 ACHIEVEMENT TEST—Cont'd

17. (2470 mg/L)(0.57 MG)(8.34 lbs/gal) = 11,742 lbs MLVSS

18. $\dfrac{(2740 \text{ mg/}L)(0.705 \text{ MGD})(8.34 \text{ lbs/gal})}{(180 \text{ mg/}L)(1.78 \text{ MGD})(8.34 \text{ lbs/gal})}$ = 6.0 days

19. $\dfrac{(2650 \text{ mg/}L)(1.38 \text{ MG})(8.34 \text{ lbs/gal}) + (1900 \text{ mg/}L)(0.117 \text{ MG})(8.34 \text{ lbs/gal})}{(5960 \text{ mg/}L)(0.075 \text{ MGD})(8.34 \text{ lbs/gal}) + (20 \text{ mg/}L)(2.9 \text{ MGD})(8.34 \text{ lbs/gal})}$

    $= \dfrac{30,499 \text{ lbs MLSS} + 1854 \text{ lbs MLSS}}{3728 \text{ lbs/day SS} + 484 \text{ lbs/day SS}} = 7.7$ days

20. $\dfrac{228 \text{ m}L/L}{772 \text{ m}L/L} = 0.30$

21. $\dfrac{3630 \text{ lbs/day}}{\dfrac{(x \text{ lbs MLSS})(71)}{100}} = 0.5$

    x = 10,225 lbs MLSS

22. $\dfrac{(2890 \text{ mg/}L)(0.485 \text{ MGD})(8.34 \text{ lbs/gal})}{(6050 \text{ mg/}L)(0.028 \text{ MGD})(8.34 \text{ lbs/gal}) + (22 \text{ mg/}L)(1.42 \text{ MGD})(8.34 \text{ lbs/gal})}$

    $= \dfrac{11,690 \text{ lbs MLSS}}{1413 \text{ lbs/day SS} + 261 \text{ lbs/day}}$

    $= 7.0$ days

23. First calculate desired lbs MLSS based on desired sludge age:

    $\dfrac{x \text{ lbs MLSS}}{3670 \text{ lbs/day SS}} = 4.8$ days

    x = 17,616 lbs MLSS

    Next, calculate actual lbs MLSS:

    (2730 mg/L)(0.77 MG)(8.34 lbs/gal) = 17,532 lbs MLSS

    Based on these calculations, no MLSS should be wasted.

24. (6340 mg/L)(x MGD)(8.34 lbs/gal) = 4100 lbs/day solids

    x = 0.078 MGD

25. $$\frac{(2870 \text{ mg/}L)(1.45 \text{ MG})(8.34 \text{ lbs/gal})}{(x \text{ lbs/day}) + (18 \text{ mg/}L)(5.68 \text{ MGD})(8.34 \text{ lbs/gal})} = 10 \text{ days}$$

$$\frac{34,707 \text{ lbs/day MLSS}}{x \text{ lbs/day} + 853 \text{ lbs/day}} = 10 \text{ days}$$

$$\frac{34,707}{10} = x + 853$$

$$3471 = x + 853$$

$$2618 \text{ lbs/day} = x$$

Chapter 13

PRACTICE PROBLEMS 13.1

1. $(245 \text{ mg/}L)(0.39 \text{ MGD})(8.34 \text{ lbs/gal}) = 797 \text{ lbs/day}$

2. $(158 \text{ mg/}L)(0.22 \text{ MGD})(8.34 \text{ lbs/gal}) = 290 \text{ lbs/day}$

3. $(221 \text{ mg/}L)(0.252 \text{ MGD})(8.34 \text{ lbs/gal}) = 464 \text{ lbs/day}$

4. $(190 \text{ mg/}L)(0.18 \text{ MGD})(8.34 \text{ lbs/gal}) = 285 \text{ lbs/day}$

PRACTICE PROBLEMS 13.2

1. $$\frac{(190 \text{ mg/}L)(0.2 \text{ MGD})(8.34 \text{ lbs/gal})}{7.5 \text{ ac}} = 42 \text{ lbs/day/ac}$$

2. $$\frac{(147 \text{ mg/}L)(0.157 \text{ MGD})(8.34 \text{ lbs/gal})}{6.5 \text{ ac}} = 30 \text{ lbs/day/ac}$$

3. $$\frac{(124 \text{ mg/}L)(0.07 \text{ MGD})(8.34 \text{ lbs/gal})}{1.9 \text{ ac}} = 38 \text{ lbs/day/ac}$$

4. $$\frac{(187 \text{ mg/}L)(x \text{ MGD})(8.34 \text{ lbs/gal})}{15 \text{ ac}} = 20 \text{ lbs/day/ac}$$

   $$x = 0.19 \text{ MGD}$$

PRACTICE PROBLEMS 13.3

1. $\dfrac{168 \text{ mg/}L \text{ Removed}}{207 \text{ mg/}L \text{ Total}}$ x 100 = 81% BOD Removed

2. $\dfrac{132 \text{ mg/}L \text{ Removed}}{262 \text{ mg/}L \text{ Total}}$ x 100 = 50% BOD Removed

3. $\dfrac{235 \text{ mg/}L \text{ Removed}}{280 \text{ mg/}L \text{ Total}}$ x 100 = 84% BOD Removed

4. $\dfrac{84 \text{ mg/}L \text{ Removed}}{140 \text{ mg/}L \text{ Total}}$ x 100 = 60% BOD Removed

PRACTICE PROBLEMS 13.4

1. First calculate hydraulic loading in ft/day:

   $$\frac{3.3 \text{ ac-ft/day}}{20 \text{ ac}} = 0.165 \text{ ft/day}$$

   Then convert to in./day:

   $$(0.165 \text{ ft/day})(12 \text{ in./ft}) = 2 \text{ in./day}$$

2. $\dfrac{5 \text{ ac-ft/day}}{15 \text{ ac}} = 0.33 \text{ ft/day}$

   $$(0.33 \text{ ft/day})(12 \text{ in./ft}) = 4 \text{ in./day}$$

3. The 18 acres must be converted to sq ft, or the gpd flow rate must be expressed as ac-ft/day. Convert gpd flow rate to ac-ft/day:

   $$\frac{2,320,000 \text{ gpd}}{(7.48 \text{ gal/cu ft})(43,560 \text{ cu ft/ac-ft})} = 7.1 \text{ ac-ft}$$

   Then calculate hydraulic loading in ft/day and in./day:

   $$\frac{7.1 \text{ ac-ft/day}}{18 \text{ ac}} = 0.39 \text{ ft/day}$$

   $$(0.39 \text{ ft/day})(12 \text{ in./ft}) = 5 \text{ in./day}$$

4. As in problem #3, either the flow rate must be converted to ac-ft/day, or the area must be expressed as sq ft. In this problem acres will be converted to sq ft:

$$(14 \text{ ac})(43{,}560 \text{ sq ft/ac}) = 609{,}840 \text{ sq ft}$$

Next calculate hydraulic loading in gpd/sq ft. Then conversion equation 1 gpm/sq ft = 1.6 in./day, may be used to convert to in./day:

$$\frac{1{,}680{,}000 \text{ gpd}}{609{,}840 \text{ sq ft}} = 2.75 \text{ gpd/sq ft}$$

Then:

$$(2.75 \text{ gpd/sq ft})\frac{(1.6 \text{ in./day})}{\text{gpd/sq ft}} = 4 \text{ in./day}$$

PRACTICE PROBLEMS 13.5

1. $\dfrac{1320 \text{ people}}{4 \text{ ac}} = 330 \text{ people/ac}$

2. $\dfrac{5460 \text{ people}}{18.5 \text{ ac}} = 295 \text{ people/ac}$

3. $\dfrac{(1680 \text{ mg}/L)(0.6 \text{ MGD})(8.34 \text{ lbs/gal})}{0.2 \text{ lbs BOD/day/person}} = 42{,}034 \text{ people}$

4. $\dfrac{(2160 \text{ mg}/L)(0.25 \text{ MGD})(8.34 \text{ lbs/gal})}{0.2 \text{ lbs BOD/day/person}} = 22{,}518 \text{ people}$

PRACTICE PROBLEMS 13.6

1. $\dfrac{17 \text{ ac-ft}}{0.42 \text{ ac-ft/day}} = 40 \text{ days}$

2. $\dfrac{(440 \text{ ft})(680 \text{ ft})(6 \text{ ft})(7.48 \text{ gal/cu ft})}{300{,}000 \text{ gpd}} = 45 \text{ days}$

3. $\dfrac{(240 \text{ ft})(390 \text{ ft})(5 \text{ ft})(7.48 \text{ gal/cu ft})}{70{,}000 \text{ gpd}} = 50 \text{ days}$

4. $\dfrac{26 \text{ ac}}{0.47 \text{ ac-ft/day}} = 55 \text{ days}$

CHAPTER 13 ACHIEVEMENT TEST

1. $(190 \text{ mg/}L)(0.36 \text{ MGD})(8.34 \text{ lbs/gal}) = 570 \text{ lbs/day}$

2. $\dfrac{(240 \text{ mg/}L)(0.28 \text{ MGD})(8.34 \text{ lbs/gal})}{8.5 \text{ ac}} = \dfrac{66 \text{ lbs/day}}{\text{ac}}$

3. $\dfrac{167 \text{ mg/}L \text{ BOD Removed}}{210 \text{ mg/}L \text{ BOD Total}} \times 100 = 80\% \text{ BOD Removed}$

4. $\dfrac{3.6 \text{ ac-ft/day}}{22 \text{ ac}} = 0.16 \text{ ft/day}$

   $(0.16 \text{ ft/day})(12 \text{ in./day}) = 2 \text{ in./day}$

5. $\dfrac{91 \text{ mg/}L \text{ Rem.}}{162 \text{ mg/}L \text{ Total}} \times 100 = 56\% \text{ BOD Removed}$

6. $(218 \text{ mg/}L)(0.288 \text{ MGD})(8.34 \text{ lbs/gal}) = 524 \text{ lbs/day}$

7. $\dfrac{(138 \text{ mg/}L)(0.075 \text{ MGD})(8.34 \text{ lbs/gal})}{\dfrac{(390 \text{ ft})(230 \text{ ft})}{43,560 \text{ sq ft/ac}}} = \dfrac{41.9 \text{ lbs/day}}{\text{ac}}$

8. $\dfrac{1,960,000 \text{ gpd}}{(19 \text{ ac})(43,560 \text{ sq ft/ac})} = 2.4 \text{ gpd/sq ft}$

   Convert gpd/sq ft to ft/day using 1 gpd/sq ft = 1.6 in./day:

   $(2.4 \text{ gpd/sq ft})\dfrac{(1.6 \text{ in./day})}{\text{gpd/sq ft}} = 4 \text{ in./day}$

9. $\dfrac{6000 \text{ people}}{20 \text{ ac}} = 300 \text{ people/ac}$

10. $\dfrac{18.2 \text{ ac-ft}}{0.51 \text{ ac-ft/day}} = 36 \text{ days}$

11. $\dfrac{(2840 \text{ mg/}L)(0.7 \text{ MGD})(8.34 \text{ lbs/gal})}{0.2 \text{ lbs BOD/day/person}} = 82,900 \text{ people}$

12. $\dfrac{(420 \text{ ft})(710 \text{ ft})(5 \text{ ft})(7.48 \text{ gal/cu ft})}{350,000 \text{ gpd}} = 32 \text{ days}$

Chapter 14

PRACTICE PROBLEMS 14.1

1. $(3.2 \text{ mg}/L)(4.4 \text{ MGD})(8.34 \text{ lbs/gal}) = 117 \text{ lbs/day}$

2. $(10 \text{ mg}/L)(1.66 \text{ MGD})(8.34 \text{ lbs/gal}) = 138 \text{ lbs/day}$

3. $(2100 \text{ mg}/L)(0.195 \text{ MGD})(8.34 \text{ lbs/gal}) = 3415 \text{ lbs/day}$

4. $(x \text{ mg}/L)(4.88 \text{ MGD})(8.34 \text{ lbs/gal}) = 312 \text{ lbs/day}$

   $x = 7.7 \text{ mg}/L$

PRACTICE PROBLEMS 14.2

1. $4.8 \text{ mg}/L + 0.9 \text{ mg}/L = 5.7 \text{ mg}/L$

2. $8.4 \text{ mg}/L = x \text{ mg}/L + 0.8 \text{ mg}/L$

   $x = 7.6 \text{ mg}/L$

3. $7.7 \text{ mg}/L + 0.5 \text{ mg}/L = 8.2 \text{ mg}/L$

4. $(9.5 \text{ mg}/L)(3.9 \text{ MGD})(8.34 \text{ lbs/gal}) = 309 \text{ lbs/day}$

PRACTICE PROBLEMS 14.3

1. $\dfrac{(10.8 \text{ mg}/L)(2.77 \text{ MGD})(8.34 \text{ lbs/gal})}{\dfrac{65}{100}} = 384 \text{ lbs/day}$

2. $\dfrac{(9.5 \text{ mg}/L)(3.52 \text{ MGD})(8.34 \text{ lbs/gal})}{\dfrac{60}{100}} = 465 \text{ lbs/day}$

3. $\dfrac{(18 \text{ mg}/L)(1.695 \text{ MGD})(8.34 \text{ lbs/gal})}{\dfrac{65}{100}} = 391 \text{ lbs/day}$

4. $\dfrac{(x \text{ mg}/L)(5.15 \text{ MGD})(8.34 \text{ lbs/gal})}{\dfrac{65}{100}} = 900 \text{ lbs/day}$

   $x = 14 \text{ mg}/L$

PRACTICE PROBLEMS 14.4

1. $\dfrac{0.6 \text{ lbs}}{(15 \text{ gal})(8.34 \text{ lbs/gal}) + 0.6 \text{ lbs}} \times 100 = 0.5\%$

2. $\dfrac{x \text{ lbs}}{(20 \text{ gal})(8.34 \text{ lbs/gal}) + x \text{ lbs}} \times 100 = 0.8$

$\dfrac{100x}{166.8 \text{ lbs} + x \text{ lbs}} = 0.8$

$100x = (0.8)\left[166.8 + x\right]$

$100x = 133.4 + 0.8x$

$99.2x = 133.4$

$x = 1.3 \text{ lbs}$

3. $1g = 0.0022 \text{ lbs}; \ 150 \text{ g} = 0.3 \text{ lbs}$

$\dfrac{0.3 \text{ lbs}}{(10 \text{ gal})(8.34 \text{ lbs/gal}) + 0.3 \text{ lbs}} \times 100 = 0.4\%$

4. $\dfrac{(10)}{100}(x \text{ lbs liq. polymer}) = \dfrac{(0.5)}{100}(167 \text{ lbs polymer sol'n})$

$x = 8.35 \text{ lbs liq. polymer}$

5. $\dfrac{(10)}{100}(x \text{ gal liq. polymer})(10.4 \text{ lbs/gal}) = \dfrac{(0.3)}{100}(50 \text{ gal polymer sol'n})(8.34 \text{ lbs/gal})$

$x = 1.2 \text{ gal liq. polymer}$

6. $\dfrac{(12)}{100}(x \text{ gal})(10.3 \text{ lbs/gal}) = \dfrac{(0.6)}{100}(100 \text{ gal})(8.34 \text{ lbs/gal})$

$x = 4 \text{ gal}$

PRACTICE PROBLEMS 14.5

1. $\dfrac{\dfrac{(10)}{100}(25 \text{ lbs}) + \dfrac{(0.5)}{100}(100 \text{ lbs})}{25 \text{ lbs} + 100 \text{ lbs}} \times 100$

$= \dfrac{2.5 \text{ lbs} + 0.5 \text{ lbs}}{125 \text{ lbs}} \times 100$

$= 2.4\% \text{ Strength}$

2. $$\frac{\dfrac{(12)}{100}(5 \text{ gal})(10.4 \text{ lbs/gal}) + \dfrac{(0.3)}{100}(32 \text{ gal})(8.4 \text{ lbs/gal})}{(5 \text{ gal})(10.4 \text{ lbs/gal}) + (32 \text{ gal})(8.4 \text{ lbs/gal})} \times 100$$

$$= \frac{6.2 \text{ lbs} + 0.8 \text{ lbs}}{52 \text{ lbs} + 268.8 \text{ lbs}} \times 100$$

$$= 2.2\% \text{ Strength}$$

3. $$\frac{\dfrac{(10)}{100}(10 \text{ gal})(10.2 \text{ lbs/gal}) + \dfrac{(0.25)}{100}(40 \text{ gal})(8.34 \text{ lbs/gal})}{(10 \text{ gal})(10.2 \text{ lbs/gal}) + (40 \text{ gal})(8.34 \text{ lbs/gal})} \times 100$$

$$= \frac{10.2 \text{ lbs} + 0.8 \text{ lbs}}{102 \text{ lbs} + 333.6 \text{ lbs}} \times 100 = 2.5\% \text{ strength}$$

4.   Use the dilution rectangle:

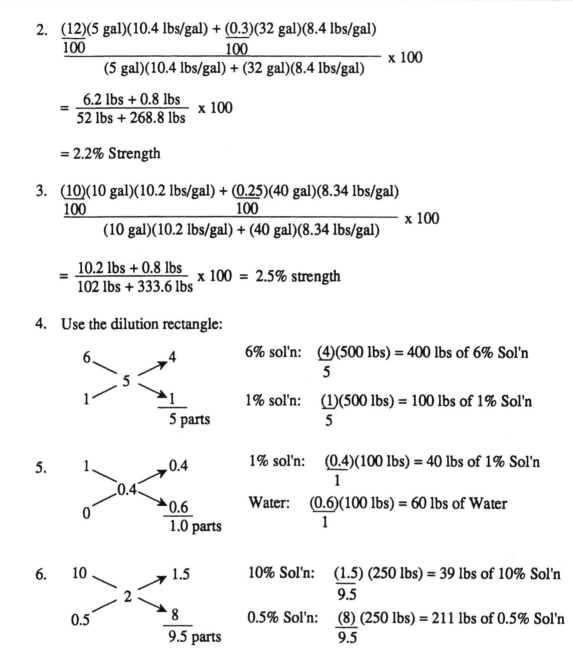

6% sol'n: $\dfrac{(4)(500 \text{ lbs})}{5} = 400$ lbs of 6% Sol'n

1% sol'n: $\dfrac{(1)(500 \text{ lbs})}{5} = 100$ lbs of 1% Sol'n

5.

1% sol'n: $\dfrac{(0.4)(100 \text{ lbs})}{1} = 40$ lbs of 1% Sol'n

Water: $\dfrac{(0.6)(100 \text{ lbs})}{1} = 60$ lbs of Water

6.

10% Sol'n: $\dfrac{(1.5)(250 \text{ lbs})}{9.5} = 39$ lbs of 10% Sol'n

0.5% Sol'n: $\dfrac{(8)(250 \text{ lbs})}{9.5} = 211$ lbs of 0.5% Sol'n

PRACTICE PROBLEM 14.6

1.   First calculate the lbs/day alum required:

$(11 \text{ mg/}L)(3.89 \text{ MGD})(8.34 \text{ lbs/gal}) = 357$ lbs/day alum req'd

Then determine gpd alum solution required:

$$\frac{357 \text{ lbs/day}}{5.36 \text{ lbs alum/gal solution}} = 67 \text{ gpd solution}$$

PRACTICE PROBLEMS 14.6—Cont'd

2. First calculate the lbs/day alum required:

   (9 mg/*L*)(1.34 MGD)(8.34 lbs/gal) = 101 lbs/day alum req'd

   Then determine gpd alum solution required:

   $$\frac{101 \text{ lbs/day}}{5.36 \text{ lbs alum/gal solution}} = 19 \text{ gpd solution}$$

3. First calculate the MGD solution required, then convert this to gpd:

   (10 mg/*L*)(2.02 MGD)(8.34 lbs/gal) = (600,000 mg/*L*)(*x* MGD)(8.34 lbs/gal)

   x = 0.0000336 MGD

   or = 34 gpd solution

   An <u>alternate approach</u> would be to calculate lbs/day solution required similar to a hypochlorite calculation:

   $$\frac{(10 \text{ mg/}L)(2.02 \text{ MGD})(8.34 \text{ lbs/gal})}{\dfrac{60}{100}} = 281 \text{ lbs/day solution}$$

   Then convert lbs/day to gpd:

   $$\frac{281 \text{ lbs/day solution}}{8.34 \text{ lbs/gal}} = 34 \text{ gpd solution}$$

4. First calculate the MGD solution required, then convert this to gpd:

   (8 mg/*L*)(4.12 MGD)(8.34 lbs/gal) = (600,000 mg/*L*)(*x* MGD)(8.34 lbs/gal)

   x = 0.0000549 MGD

   or = 54.9 gpd

PRACTICE PROBLEMS 14.7

1. $\dfrac{25 \text{ gpm}}{90 \text{ gpm}} \times 100 = 28\%$

2. $\dfrac{20 \text{ gpm}}{85 \text{ gpm}} \times 100 = 24\%$

3. $\dfrac{15 \text{ gpm}}{80 \text{ gpm}} \times 100 = 19\%$

4. $\dfrac{30 \text{ gpm}}{100 \text{ gpm}} \times 100 = 30\%$

PRACTICE PROBLEMS 14.8

1. $$\frac{(40\ \underline{\text{gal}})(3785\ \underline{\text{mL}})}{\text{day}\qquad\text{gal}} = 105\ \text{mL/min}$$

2. $$\frac{(35\ \underline{\text{gal}})(3785\ \underline{\text{mL}})}{\text{day}\qquad\text{gal}} = 92\ \text{mL/min}$$

3. First calculate the gpd setting required:

    $(8\ \text{mg/L})(0.89\ \text{MGD})(8.34\ \text{lbs/gal}) = (600{,}000\ \text{mg/L})(x\ \text{MGD})(8.34\ \text{lbs/gal})$

    $x = 0.0000118\ \text{MGD}$

    or $= 11.8\ \text{gpd}$

    Then convert gpd to ml/min flow rate:

    $$\frac{(11.8\ \underline{\text{gal}})(3785\ \underline{\text{mL}})}{\text{day}\qquad\text{gal}} = 31\ \text{mL/min}$$

4. First calculate the gpd setting required:

    $(10\ \text{mg/L})(1.24\ \text{MGD})(8.34\ \text{lbs/gal}) = (550{,}000\ \text{mg/L})(x\ \text{MGD})(8.34\ \text{lbs/gal})$

    $x = 0.0000225\ \text{MGD}$

    or $= 22.5\ \text{gpd}$

    Then convert gpd to ml/min flow rate:

    $$\frac{(22.5\ \underline{\text{gal}})(3785\ \underline{\text{mL}})}{\text{day}\qquad\text{gal}} = 59\ \text{mL/min}$$

PRACTICE PROBLEMS 14.9

1. First calculate lbs/min feed rate:

    $$\frac{1.8\ \text{lbs}}{30\ \text{min}} = 0.06\ \text{lbs/min}$$

    Then convert lbs/min to lbs/day feed rate:

    $$(0.06\ \underline{\text{lbs}})(1440\ \underline{\text{min}}) = 86\ \text{lbs/day}$$
    $$\qquad\ \text{min}\qquad\text{day}$$

PRACTICE PROBLEMS 14.9—Cont'd

2. $\dfrac{1.375 \text{ lbs}}{25 \text{ min}}$ = 0.055 lbs/min

   Then (0.055 lbs/min)(1440 min/day) = 79 lbs/day

3. 10 oz = 0.63 lbs

   Determine lbs of chemical used:

   1.95 lbs bucket + chemical
   − 0.63 lbs bucket
   ‾‾‾‾‾‾‾‾‾‾‾‾‾‾‾‾‾‾
   1.32 lbs chemical

   Calculate lbs/min feed rate:

   $\dfrac{1.32 \text{ lbs chemical}}{30 \text{ min}}$ = 0.044 lbs/min

   Then convert lbs/min feed rate to lbs/day feed rate:

   (0.044 lbs/min)(1440 min/day) = 63 lbs/day

4. The lbs chemical used is:

   2.3 lbs chemical + bucket
   − 0.4 lbs bucket
   ‾‾‾‾‾‾‾‾‾‾‾‾‾‾‾‾
   1.9 lbs chemical

   Next, calculate lbs/min feed rate:

   $\dfrac{1.9 \text{ lbs chemical}}{30 \text{ min}}$ = 0.063 lbs/min

   Then convert lbs/min feed rate to lbs/day feed rate:

   (0.063 lbs/min)(1440 min/day) = 91 lbs/day

PRACTICE PROBLEMS 14.10

1. First calculate the mL/min solution feed rate:

   $\dfrac{720 \text{ mL}}{5 \text{ min}}$ = 144 mL/min

   Then convert mL/min feed rate to gpd feed rate:

   $\dfrac{(144 \text{ mL})}{\text{min}} \dfrac{(1 \text{ gal})}{3785 \text{ mL}} \dfrac{(1440 \text{ min})}{\text{day}}$ = 55 gpd Solution

   Now the lbs/day feed can be calculated:

   (13,000 mg/L)(0.000055 MGD)(8.34 lbs/gal) = 6.0 lbs/day polymer

2.   First calculate the m*L*/min solution feed rate:

$$\frac{950 \text{ m}L}{5 \text{ min}} = 190 \text{ m}L/\text{min}$$

Then convert m*L*/min feed rate to gpd feed rate:

$$\frac{(190 \text{ m}L)}{\text{min}} \frac{(1 L)}{1000 \text{ m}L} \frac{(1 \text{ gal})}{3.785 L} \frac{(1440 \text{ min})}{\text{day}} = 72 \text{ gpd Solution}$$

The lbs/day polymer feed can now be calculated:

(14,000 mg/*L*)(0.000072 MGD)(8.34 lbs/gal) = 8.4 lbs/day

3.   The m*L*/min solution feed rate is:

$$\frac{600 \text{ m}L}{5 \text{ min}} = 120 \text{ m}L/\text{min}$$

The m*L*/min solution feed rate can be converted to gpd feed rate:

$$\frac{(120 \text{ m}L)}{\text{min}} \frac{(1 \text{ gal})}{3785 L} \frac{(1440 \text{ min})}{\text{day}} = 46 \text{ gpd Solution}$$

The polymer dosage rate in lbs/day can now be determined:

(11,000 mg/*L*)(0.000046 MGD)(8.34 lbs/gal)(1.1 sq. gr.) = 4.6 lbs/day

4.   The m*L*/min solution feed rate is:

$$\frac{810 \text{ m}L}{5 \text{ min}} = 162 \text{ m}L/\text{min}$$

The m*L*/min solution feed can be converted to gpd feed rate:

$$\frac{(162 \text{ m}L)}{\text{min}} \frac{(1 \text{ gal})}{3785 L} \frac{(1440 \text{ min})}{\text{day}} = 62 \text{ gpd Solution}$$

The polymer dosage rate in lbs/day is:

(5,000 mg/*L*)(0.000062 MGD)(8.34 lbs/gal)(1.04 sp. gr.) = 2.7 lbs/day

PRACTICE PROBLEM 14.11

1. $$\frac{(0.785)(3 \text{ ft})(3 \text{ ft})(1.3 \text{ ft})(7.48 \text{ gal/cu ft})}{3 \text{ min}} = 23 \text{ gpm}$$

2. $$\frac{(0.785)(4 \text{ ft})(4 \text{ ft})(1.2 \text{ ft})(7.48 \text{ gal/cu ft})}{5 \text{ min}} = 23 \text{ gpm}$$

3. $$\frac{(0.785)(5 \text{ ft})(5 \text{ ft})(x \text{ ft})(7.48 \text{ gal/cu ft})}{4 \text{ min}} = 25 \text{ gpm}$$

$$x = 0.7 \text{ ft}$$

PRACTICE PROBLEMS 14.11—Cont'd

4. $\dfrac{(0.785)(4\text{ ft})(4\text{ ft})(1.4\text{ ft})(7.48\text{ gal/cu ft})}{3\text{ min}} = 44\text{ gpm}$

PRACTICE 14.12

1. $\dfrac{530\text{ lbs}}{7\text{ days}} = 76\text{ lbs/day average}$

2. $\dfrac{2100\text{ lbs}}{95\text{ lbs/day}} = 22.1\text{ days}$

3. $\dfrac{921\text{ lbs}}{60\text{ lbs/day}} = 15.4\text{ days}$

4. First calcualte the gallons supply:

    $(0.785)(4\text{ ft})(4\text{ ft})(3.2\text{ ft})(7.48\text{ gal/cu ft}) = 301\text{ gal}$

    Then calculate day's supply:

    $\dfrac{301\text{ gal}}{92\text{ gpd}} = 3.3\text{ days}$

CHAPTER 14 ACHIEVEMENT TEST

1. $(9\text{ mg/L})(3.175\text{ MGD})(8.34\text{ lbs/gal}) = 238\text{ lbs/day}$

2. $(6.9\text{ mg/L})(3.12\text{ MGD})(8.34\text{ lbs/gal}) = 276\text{ lbs/day}$
    $$\dfrac{65}{100}$$

3. $\dfrac{x\text{ lbs}}{(30\text{ gal})(8.34\text{ lbs/gal}) + x\text{ lbs}} \times 100 = 0.1$

    $\dfrac{100\,x}{250.2\text{ lbs} + x\text{ lbs}} = 0.1$

    $100\,x = (0.1)(250.2 + x)$

    $100\,x = 25.02 + 0.1\,x$

    $100\,x - 0.1x = 25.02$

    $99.9\,x = 25.02$

    $x = 0.25\text{ lbs}$

4.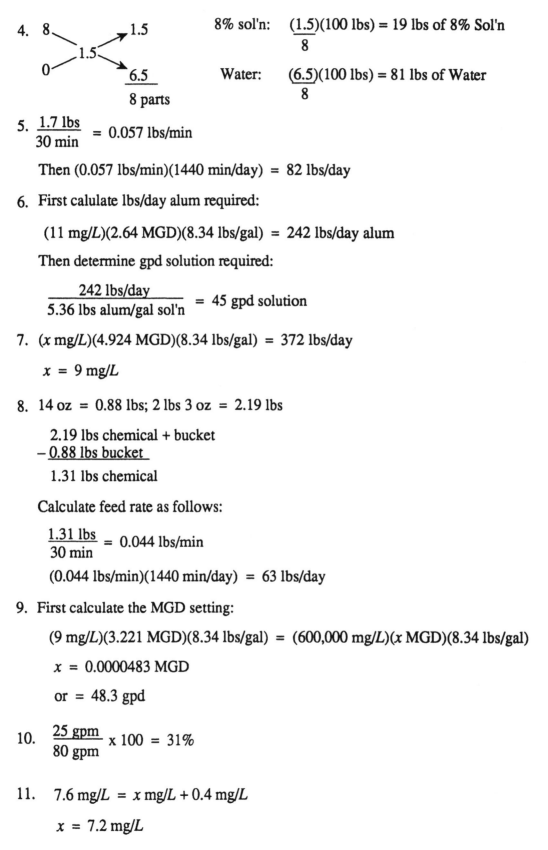

8% sol'n: $\dfrac{(1.5)(100 \text{ lbs})}{8} = 19$ lbs of 8% Sol'n

Water: $\dfrac{(6.5)(100 \text{ lbs})}{8} = 81$ lbs of Water

5. $\dfrac{1.7 \text{ lbs}}{30 \text{ min}} = 0.057$ lbs/min

Then (0.057 lbs/min)(1440 min/day) = 82 lbs/day

6. First calulate lbs/day alum required:

(11 mg/L)(2.64 MGD)(8.34 lbs/gal) = 242 lbs/day alum

Then determine gpd solution required:

$\dfrac{242 \text{ lbs/day}}{5.36 \text{ lbs alum/gal sol'n}} = 45$ gpd solution

7. (x mg/L)(4.924 MGD)(8.34 lbs/gal) = 372 lbs/day

$x = 9$ mg/L

8. 14 oz = 0.88 lbs; 2 lbs 3 oz = 2.19 lbs

$\begin{array}{l} 2.19 \text{ lbs chemical + bucket} \\ \underline{- \ 0.88 \text{ lbs bucket}} \\ 1.31 \text{ lbs chemical} \end{array}$

Calculate feed rate as follows:

$\dfrac{1.31 \text{ lbs}}{30 \text{ min}} = 0.044$ lbs/min

(0.044 lbs/min)(1440 min/day) = 63 lbs/day

9. First calculate the MGD setting:

(9 mg/L)(3.221 MGD)(8.34 lbs/gal) = (600,000 mg/L)(x MGD)(8.34 lbs/gal)

$x = 0.0000483$ MGD

or = 48.3 gpd

10. $\dfrac{25 \text{ gpm}}{80 \text{ gpm}} \times 100 = 31\%$

11. 7.6 mg/L = x mg/L + 0.4 mg/L

$x = 7.2$ mg/L

ACHIEVEMENT TEST—Cont'd

12. $\dfrac{(10)(x \text{ gal})(9.8 \text{ lbs/gal})}{100} = \dfrac{(0.5)(50 \text{ gal})(8.34 \text{ lbs/gal})}{100}$

$x = 2.1 \text{ gal}$

13. First calculate the mL/min feed rate:

$\dfrac{640 \text{ m}L}{5 \text{ min}} = 128 \text{ m}L/\text{min}$

Then convert mL/min to gpd feed rate:

$(128 \dfrac{\text{m}L}{\text{min}})(\dfrac{1 \text{ gal}}{3785 \, L})(\dfrac{1440 \text{ min}}{\text{day}}) = 48.7 \text{ gpd}$

Now the polymer dosage in lbs/day may be determined:

$(11,000 \text{ mg}/L)(0.0000487 \text{ MGD})(8.34 \text{ lbs/gal}) = 4.5 \text{ lbs/day polymer}$

14. $\dfrac{(0.785)(5 \text{ ft})(5 \text{ ft})(x \text{ ft})(7.48 \text{ gal/cu ft})}{5 \text{ min}} = 25 \text{ gpm}$

$x = 0.9 \text{ ft}$

15. $\dfrac{15 \text{ gpm}}{80 \text{ gpm}} \times 100 = 19\%$

16. $(9.4 \text{ mg}/L)(4.1 \text{ MGD})(8.34 \text{ lbs/gal}) = 321 \text{ lbs/day}$

17. $\dfrac{1960 \text{ lbs}}{80 \text{ lbs/day}} = 24.5 \text{ days}$

18. $(\dfrac{40 \text{ gal}}{\text{day}})(\dfrac{3785 \text{ m}L}{\text{gal}})(\dfrac{1 \text{ day}}{1440 \text{ min}}) = 105 \text{ m}L/\text{min}$

19. $\dfrac{810 \text{ m}L}{5 \text{ min}} = 162 \text{ m}L/\text{min}$

Convert mL/min to gpd feed rate:

$(162 \dfrac{\text{m}L}{\text{min}})(\dfrac{1 \text{ gal}}{3785 \, L})(\dfrac{1440 \text{ min}}{\text{day}}) = 62 \text{ gpd}$

Then determine polymer dosage rate in lbs/day:

$(8,000 \text{ mg}/L)(0.000062 \text{ MGD})(8.34 \text{ lbs/gal}) = 4.1 \text{ lbs/day polymer}$

20. First determine MGD then gpd setting required:

    $(8 \text{ mg/}L)(3.142 \text{ MGD})(8.34 \text{ lbs/gal}) = (550{,}000 \text{ mg/}L)(x \text{ MGD})(8.34 \text{ lbs/gal})$

    $x = 0.0000457 \text{ MGD}$

    or $= 46 \text{ gpd}$

    Then convert gpd to mL/min feed rate:

    $$\frac{(46 \text{ gal})}{\text{day}} \frac{(3785 \text{ mL})}{\text{gal}} \frac{(1 \text{ day})}{1440 \text{ min}} = 121 \text{ mL/min}$$

21. $$\frac{(0.785)(3 \text{ ft})(3 \text{ ft})(1.4 \text{ ft})(7.48 \text{ gal/cu ft})}{3 \text{ min}} = 25 \text{ gpm}$$

22. $$\frac{(10.4 \text{ mg/}L)(2.942 \text{ MGD})(8.34 \text{ lbs/gal})}{\dfrac{65}{100}} = 393 \text{ lbs/day hypochlorite}$$

23.

    15   0.5   15% sol'n: $\dfrac{(0.5)(100 \text{ lbs})}{14.5} = 3$ lbs of 15% Sol'n

                1

    0.5   14   0.5% Sol'n: $\dfrac{(14)(100 \text{ lbs})}{14.5} = 97$ lbs of 0.5% Sol'n

                14.5 parts

24. $$\frac{\dfrac{(10)}{100}(5 \text{ gal})(10.1 \text{ lbs/gal}) + \dfrac{(0.3)}{100}(20 \text{ gal})(8.34 \text{ lbs/gal})}{(5 \text{ gal})(10.1 \text{ lbs/gal}) + (20 \text{ gal})(8.34 \text{ lbs/gal})} \times 100$$

    $$= \frac{5.1 \text{ lbs} + 0.5 \text{ lbs}}{50.5 + 166.8} \times 100$$

    $$= \frac{5.6}{217.3} \times 100$$

    $$= 2.6\% \text{ strength}$$

Chapter 15

PRACTICE PROBLEMS 15.1

1. $(155 \text{ mg/}L)(4.73 \text{ MGD})(8.34 \text{ lbs/gal}) = 6114 \text{ lbs/day solids}$

2. $(177 \text{ mg/}L \text{ Rem.})(3.6 \text{ MGD})(8.34 \text{ lbs/gal}) = 5314 \text{ lbs/day solids}$

PRACTICE PROBLEMS 15.1—Cont'd

3. First calculate BOD removed:

    (118 mg/*L* BOD Rem.)(1.9 MGD)(8.34 lbs/gal) = 1870 lbs/day BOD Removed

    Use the *Y*-value to determine dry solids produced:

    $$\frac{0.5 \text{ lbs/day SS}}{1 \text{ lb/day BOD Removed}} = \frac{x \text{ lbs/day SS}}{1870 \text{ lbs/day BOD Removed}}$$

    $$(0.5)(1870) = x$$

    935 lbs/day dry solids = *x*

4. First calculate BOD removed:

    (158 mg/*L* BOD Rem.)(2.67 MGD)(8.34 lbs/gal) = 3518 lbs/day BOD Removed

    Use the *Y*-value to determine dry solids produced:

    $$\frac{0.45 \text{ lbs/day SS}}{1 \text{ lb/day BOD Removed}} = \frac{x \text{ lbs/day SS}}{3518 \text{ lbs/day BOD Removed}}$$

    $$(0.45)(3518) = x$$

    1583 lbs/day dry solids = *x*

PRACTICE PROBLEMS 15.2

1. $x = \dfrac{0.56 \text{ grams}}{28 \text{ grams}} \times 100$

    $x = 2\%$ solids

2. $4.1 = \dfrac{8{,}300 \text{ lbs/day solids}}{x \text{ lbs/day sludge}} \times 100$

    $x = 202{,}439$ lbs/day sludge

3. $5.3 = \dfrac{x \text{ lbs/day solids}}{(9200 \text{ gal sludge})(8.34 \text{ lbs/gal})} \times 100$

    $x = 4067$ lbs/day solids

4. $5.1 = \dfrac{1480 \text{ lbs/day solids}}{(x \text{ gpd sludge})(8.34 \text{ lbs/gal})} \times 100$

    $x = 3480$ gpd sludge

5. $4.6 = \dfrac{920 \text{ lbs/day solids}}{(x \text{ gpd sludge})(8.34 \text{ lbs/gal})} \times 100$

    $x = 2398$ gpd sludge

PRACTICE PROBLEMS 15.3

1. $(21,500 \text{ lbs/day})(0.04) = (x \text{ lbs/day})(0.065)$

    $$\frac{(21,500)(0.04)}{0.065} = x$$

    $13,231 \text{ lbs/day} = x$
    Thickened Sludge

2. $(2850 \text{ gpd})(8.34 \text{ lbs/gal})(0.05) = (x \text{ gpd})(8.61 \text{ lbs/gal})(0.07)$

    $$\frac{(2850)(8.34)(0.05)}{(8.61)(0.07)} = x$$

    $1972 \text{ gpd} = x$
    Thickened Sludge

3. $(12,600 \text{ lbs/day})(0.035) = (x \text{ lbs/day})(0.055)$

    $$\frac{(12,600)(0.035)}{0.055} = x$$

    $8018 \text{ lbs/day} = x$
    Thickened Sludge

4. $(5600 \text{ gpd})(8.34 \text{ lbs/gal})(0.038) = (x \text{ gpd})(8.58 \text{ lbs/gal})(0.063)$

    $$\frac{(5600)(8.34)(0.038)}{(8.58)(0.063)} = x$$

    $3283 \text{ gpd} = x$
    Thickened Sludge

PRACTICE PROBLEMS 15.4

1. $$\frac{(60 \text{ gpm} + 85 \text{ gpm})(1440 \text{ min/day})}{(0.785)(25 \text{ ft})(25 \text{ ft})} = \frac{(145 \text{ gpm})(1440 \text{ min/day})}{491 \text{ sq ft}}$$

    $$= 411 \text{ gpd/sq ft}$$

2. $$\frac{(165 \text{ gpm})(1440 \text{ min/day})}{(0.785)(25 \text{ ft})(25 \text{ ft})} = 484 \text{ gpd/sq ft}$$

3. $$\frac{(110,000 \text{ gpd})(8.34 \text{ lbs/gal})(3.8)}{(0.785)(40 \text{ ft})(40 \text{ ft})} \bigg/ 100 = \frac{28 \text{ lbs/day}}{\text{sq ft}}$$

PRACTICE PROBLEMS 15.4—Cont'd

4. $$\frac{(50 \text{ gpm})(1440 \frac{\text{min}}{\text{day}})(8.34 \frac{\text{lbs}}{\text{gal}})(\frac{3.6}{100})}{(0.785)(30 \text{ ft})(30 \text{ ft})} = 31 \frac{\text{lbs/day}}{\text{sq ft}}$$

5. $$\frac{(0.785)(45 \text{ ft})(45 \text{ ft})(3.5 \text{ ft})(7.48 \text{ gal/cu ft})}{(25 \text{ gpm})(1440 \text{ min/day})} = 1 \text{ day}$$

6. $$\frac{(0.785)(35 \text{ ft})(35 \text{ ft})(4.1 \text{ ft})(7.48 \text{ gal/cu ft})}{(28 \text{ gpm})(60 \text{ min/hr})} = 17.6 \text{ hrs}$$

7. $3\% = 30,000 \text{ mg/L}$

   $0.8\% = 8,000 \text{ mg/L}$

   $$\frac{22,000 \text{ mg/L Rem.}}{30,000 \text{ mg/L}} \times 100 = 73\%$$

8. $$\frac{2.6\% \text{ Removed}}{3.3\% \text{ Entering}} \times 100 = 79\%$$

9. $$\frac{8.1}{3.2} = 2.5$$

10. $$\frac{7.8}{2.9} = 2.7$$

11. First determine the lbs/day solids entering the thickener:

    $$(120 \text{ gpm})(1440 \text{ min/day})(8.34 \text{ lbs/gal})(\frac{3.2}{100}) = 46,117 \text{ lbs/day}$$

    Then calculate the lbs/day solids leaving the thickener via the underflow:

    $$(45 \text{ gpm})(1440 \text{ min/day})(8.34 \text{ lbs/gal})(\frac{7.8}{100}) = 42,154 \text{ lbs/day}$$

    Next, calculate the lbs/day solids leaving the thickener via the effluent flow:

    $$(120 \text{ gpm} - 45 \text{ gpm})(1440 \text{ min/day})(8.34 \text{ lbs/gal})(\frac{0.058}{100}) = 522 \text{ lbs/day}$$

    To summarize:

    Solids Entering = 46,117 lbs/day

    Solids Leaving = 42,676 lbs/day
    (42,154 lbs/day + 522 lbs/day)

    Since more solids are entering than are leaving, the sludge blanket will <u>rise</u>.

12. a) First determine the lbs/day solids entering the thickener:

$$\frac{(105 \text{ gpm})(1440 \text{ min/day})(8.34 \text{ lbs/gal})(\underline{3.4})}{100} = 42{,}874 \text{ lbs/day}$$

Then calculate the lbs/day solids leaving the thickener:

<u>Solids Leaving via Underflow</u>

$$\frac{(55 \text{ gpm})(1440 \text{ min/day})(8.34 \text{ lbs/gal})(\underline{6.9})}{100} = 45{,}576 \text{ lbs/day}$$

<u>Solids Leaving via Effluent Flow</u>

$$\frac{(105 \text{ gpm} - 55 \text{ gpm})(1440 \text{ min/day})(8.34 \text{ lbs/gal})(\underline{0.051})}{100} = 306 \text{ lbs/day}$$

To summarize:

Solids Entering = 42,874 lbs/day

Solids Leaving = 45,882 lbs/day
(45,576 lbs/day + 306 lbs/day)

Since more solids are leaving than are entering, the sludge blanket will <u>drop</u>.

b) The lbs/day drop is

$$\begin{array}{r} 45{,}882 \text{ lbs/day} \\ -\,\underline{42{,}874 \text{ lbs/day}} \\ 3008 \text{ lbs/day} \end{array}$$

13. First convert lbs/day storage rate to lbs/hr:

$$\frac{9000 \text{ lbs/day}}{24 \text{ hrs/day}} = 375 \text{ lbs/hr}$$

Then calculate rise time:

$$\text{Fill Time, hrs} = \frac{(0.785)(25 \text{ ft})(25 \text{ ft})(1.5 \text{ ft})(7.48 \frac{\text{gal}}{\text{cu ft}})(8.34 \frac{\text{lbs}}{\text{gal}})(\frac{6.5}{100})}{375 \text{ lbs/hr}}$$

$$= 8.0 \text{ hrs}$$

14. First convert lbs/day storage rate to lbs/hr:

$$\frac{12{,}000 \text{ lbs/day}}{24 \text{ hrs/day}} = 500 \text{ lbs/hr}$$

Next, calculate rise time:

$$\text{Fill Time, hrs} = \frac{(0.785)(25 \text{ ft})(25 \text{ ft})(2.3 \text{ ft})(7.48 \text{ gal/cu ft})(8.34 \text{ lbs/gal})(\frac{7}{100})}{500 \text{ lbs/hr}}$$

$$= 9.9 \text{ hrs}$$

PRACTICE PROBLEMS 15.4—Cont'd

15. First calculate the desired solids storage rate:

$$\frac{2.8 \text{ ft}}{5 \text{ ft}} = \frac{x \text{ lbs/min}}{50 \text{ lbs/min}}$$

$$\frac{(2.8)(50)}{5} = x$$

$$28 \text{ lbs/min} = x$$
Solids Storage Rate

Next, calculate the solids withdrawal rate required to achieve the 28 lbs/min storage rate:

$$50 \text{ lbs/min} = \underset{\text{withdrawal}}{x \text{ lbs/min}} + \underset{\text{storage}}{28 \text{ lbs/min}}$$

$$22 \text{ lbs/min} = x$$
Solids Withdrawal

Then determine the gpm sludge withdrawal rate necessary for a 22 lbs/min solids withdrawal:

$$\underset{\text{Sl. Withd.}}{(x \text{ gpm})} \ \underset{\text{lbs/gal}}{(8.34)} \ \frac{(5.8)}{100} = 22 \text{ lbs/min}$$

$$x = \frac{22}{(8.34)(0.058)}$$

$$x = 45 \text{ gpm}$$
Sludge Withdrawal

16. First calculate the desired solids storage rate:

$$\frac{3.1 \text{ ft}}{6 \text{ ft}} = \frac{x \text{ lbs/min}}{58 \text{ lbs/min}}$$

$$\frac{(3.1)(58)}{6} = x$$

$$30 \text{ lbs/min} = x$$
Solids Storage Rate

Next, calculate the solids withdrawal rate required to achieve the 30 lbs/min storage rate:

$$58 \text{ lbs/min} = \underset{\text{withdrawal}}{x \text{ lbs/min}} + \underset{\text{storage}}{30 \text{ lbs/min}}$$

$$28 \text{ lbs/min} = x$$
Solids Withdrawal

Then the gpm sludge withdrawal rate can be calculated:

$$\underset{\text{Sl. Withd.}}{(x \text{ gpm})} \ \underset{\text{lbs/gal}}{(8.34)} \ \frac{(5.5)}{100} = 28 \text{ lbs/min}$$

$$x = \frac{28}{(8.34)(0.055)}$$

$$x = 61 \text{ gpm}$$
Sludge Withdrawal

PRACTICE PROBLEMS 15.5

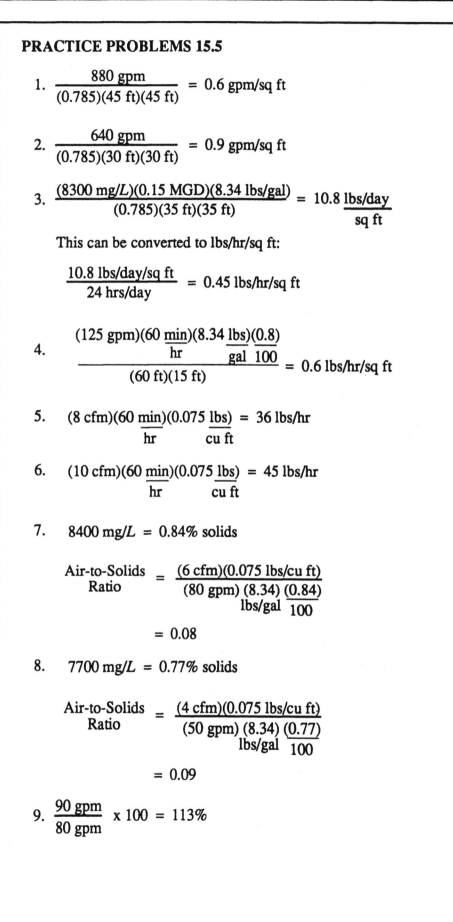

1. $\dfrac{880 \text{ gpm}}{(0.785)(45 \text{ ft})(45 \text{ ft})} = 0.6 \text{ gpm/sq ft}$

2. $\dfrac{640 \text{ gpm}}{(0.785)(30 \text{ ft})(30 \text{ ft})} = 0.9 \text{ gpm/sq ft}$

3. $\dfrac{(8300 \text{ mg/}L)(0.15 \text{ MGD})(8.34 \text{ lbs/gal})}{(0.785)(35 \text{ ft})(35 \text{ ft})} = 10.8 \dfrac{\text{lbs/day}}{\text{sq ft}}$

   This can be converted to lbs/hr/sq ft:

   $\dfrac{10.8 \text{ lbs/day/sq ft}}{24 \text{ hrs/day}} = 0.45 \text{ lbs/hr/sq ft}$

4. $\dfrac{(125 \text{ gpm})(60 \dfrac{\text{min}}{\text{hr}})(8.34 \dfrac{\text{lbs}}{\text{gal}})(\dfrac{0.8}{100})}{(60 \text{ ft})(15 \text{ ft})} = 0.6 \text{ lbs/hr/sq ft}$

5. $(8 \text{ cfm})(60 \dfrac{\text{min}}{\text{hr}})(0.075 \dfrac{\text{lbs}}{\text{cu ft}}) = 36 \text{ lbs/hr}$

6. $(10 \text{ cfm})(60 \dfrac{\text{min}}{\text{hr}})(0.075 \dfrac{\text{lbs}}{\text{cu ft}}) = 45 \text{ lbs/hr}$

7. $8400 \text{ mg/}L = 0.84\% \text{ solids}$

   $\dfrac{\text{Air-to-Solids}}{\text{Ratio}} = \dfrac{(6 \text{ cfm})(0.075 \text{ lbs/cu ft})}{(80 \text{ gpm}) (8.34 \dfrac{\text{lbs/gal}}{}) (\dfrac{0.84}{100})}$

   $= 0.08$

8. $7700 \text{ mg/}L = 0.77\% \text{ solids}$

   $\dfrac{\text{Air-to-Solids}}{\text{Ratio}} = \dfrac{(4 \text{ cfm})(0.075 \text{ lbs/cu ft})}{(50 \text{ gpm}) (8.34 \dfrac{\text{lbs/gal}}{}) (\dfrac{0.77}{100})}$

   $= 0.09$

9. $\dfrac{90 \text{ gpm}}{80 \text{ gpm}} \times 100 = 113\%$

PRACTICE PROBLEMS 15.5—Cont'd

10. $110 = \dfrac{x \text{ gpm}}{60 \text{ gpm}} \times 100$

$\dfrac{(110)(60)}{100} = x$

$66 \text{ gpm} = x$

11. $\dfrac{7370 \text{ mg/L Solids Removed}}{7600 \text{ mg/L in Influent}} \times 100 = 97\%$

12. $\dfrac{4.6}{0.825} = 5.6$

PRACTICE PROBLEMS 15.6

1. $(35 \text{ gpm})(60 \dfrac{\text{min}}{\text{hr}}) = 2100 \text{ gph}$

2. $\dfrac{85{,}400 \text{ gpd}}{24 \text{ hrs/day}} = 3558 \text{ gph}$

3. $(65 \text{ gpm})(60 \dfrac{\text{min}}{\text{hr}})(\dfrac{20 \text{ min}}{21.5 \text{ min}}) = 3628 \text{ gph}$

4. $\dfrac{(76{,}000 \text{ gpd})}{24 \text{ hrs/day}} \dfrac{(23 \text{ min})}{25 \text{ min.}} = 2913 \text{ gph}$

5. $7500 \text{ mg/L} = 0.75\%$

$\dfrac{(105{,}000 \text{ gpd})(8.34 \text{ lbs/gal})(\dfrac{0.75}{100})}{24 \text{ hrs/day}} = 274 \text{ lbs/hr}$

6. $(70 \text{ gpm})(60 \dfrac{\text{min}}{\text{hr}})(8.34 \text{ lbs/gal})(\dfrac{0.74}{100})(\dfrac{25 \text{ min}}{27 \text{ min}}) = 240 \text{ lbs/hr}$

7. $\dfrac{(30 \text{ cu ft})(7.48 \text{ gal/cu ft})(8.34 \text{ lbs/gal})(\dfrac{6.8}{100})}{(70 \text{ gpm})(8.34 \text{ lbs/gal})(\dfrac{0.73}{100})} = 30 \text{ min}$

8. $\dfrac{(20 \text{ cu ft})(7.48 \text{ gal/cu ft})(8.34 \text{ lbs/gal})(\dfrac{8}{100})}{(50 \text{ gpm})(8.34 \text{ lbs/gal})(\dfrac{0.74}{100})} = 32 \text{ min}$

9. $\dfrac{7100 \text{ mg/}L}{7800 \text{ mg/}L} \times 100 = 91\%$

10. $\dfrac{0.65\%}{0.9\%} = 72\%$

11. $\dfrac{(15 \text{ cu ft}) \; (62.4) \atop \text{lbs/cu ft} \; \dfrac{(4.2)}{100} + (3.5 \text{ cu ft}) \; (62.4) \atop \text{lbs/cu ft} \; \dfrac{(7.5)}{100}}{(15 \text{ cu ft})(62.4 \text{ lbs/cu ft}) + (3.5 \text{ cu ft})(62.4 \text{ lbs/cu ft})} \times 100$

$= \dfrac{39.3 \text{ lbs} + 16.4 \text{ lbs}}{936 \text{ lbs} + 218.4 \text{ lbs}} \times 100$

$= 4.8\%$

12. $\dfrac{(11 \text{ cu ft}) (62.4 \text{ lbs/cu ft}) \dfrac{(3.5)}{100} + (3 \text{ cu ft}) (62.4 \text{ lbs/cu ft}) \dfrac{(7.8)}{100}}{(11 \text{ cu ft})(62.4 \text{ lbs/cu ft}) + (3 \text{ cu ft})(62.4 \text{ lbs/cu ft})} \times 100$

$= \dfrac{24.0 \text{ lbs} + 14.6 \text{ lbs}}{686.4 \text{ lbs} + 187.2 \text{ lbs}} \times 100$

$= 4.4\%$

CHAPTER 15—ACHIEVEMENT TEST

1. $\dfrac{0.58 \text{ grams}}{30 \text{ grams}} \times 100 = 1.9\%$

2. $(176 \text{ mg/}L \text{ Removed})(3.2 \text{ MGD})(8.34 \text{ lbs/gal}) = 4697 \text{ lbs/day SS}$

3. $(3700 \text{ gpd})(8.34 \text{ lbs/gal})\dfrac{(3.8)}{100} = (x \text{ gpd})(8.34 \text{ lbs/gal})\dfrac{(7)}{100}$

$\dfrac{(3700)(\cancel{8.34})(0.038)}{(\cancel{8.34})(0.07)} = x$

$2009 \text{ gpd} = x$

4. $(9400 \text{ gal})(8.34 \text{ lbs/gal})\dfrac{(4.7)}{100} = 3685 \text{ lbs/day solids}$

CHAPTER 15—ACHIEVEMENT TEST—Cont'd

5. (125 mg/*L* BOD Rem.)(2.92 MGD)(8.34 lbs/gal) = 3044 lbs/day BOD Removed

Use the Y-value to estimate sludge produced:

$$\frac{0.5 \text{ lbs SS}}{1 \text{ lb BOD Rem.}} = \frac{x \text{ lbs/day SS}}{3044 \text{ lbs/day BOD Rem.}}$$

(0.5)(3044) = *x*

1522 lbs/day = *x*
SS Produced

6. $\dfrac{(0.785)(40 \text{ ft})(40 \text{ ft})(4 \text{ ft})(7.48 \text{ gal/cu ft})}{(30 \text{ gpm})(60 \text{ min/hr})}$ = 20.9 hrs

7. $\dfrac{7.5}{2.9}$ = 2.6

8. $\dfrac{(7850 \text{ mg/}L)(0.14 \text{ MGD})(8.34 \text{ lbs/gal})}{(0.785)(35 \text{ ft})(35 \text{ ft})}$ = 9.5 lbs/day/sq ft

Convert lbs/day/sq ft to lbs/hr/sq ft:

$\dfrac{9.5 \text{ lbs/day/sq ft}}{24 \text{ hrs/day}}$ = 0.4 lbs/day/sq ft

9. $\dfrac{6680 \text{ mg/}L}{6900 \text{ mg/}L}$ x 100 = 97%

10. $\dfrac{(160 \text{ gpm})(1440 \text{ min/day})}{(0.785)(25 \text{ ft})(25 \text{ ft})}$ = 470 gpd/sq ft

11. $\dfrac{2.8}{3.1}$ x 100 = 90%

12. (8 cfm)(60 $\dfrac{\text{min}}{\text{hr}}$)(0.075 $\dfrac{\text{lbs}}{\text{cu ft}}$) = 36 lbs/hr

13. a) First calculate lbs/day solids entering the thickener:

$$(105 \text{ gpm})(1440 \text{ min/day})(8.34 \text{ lbs/gal})\left(\frac{3}{100}\right) = 37{,}830 \text{ lbs/day}$$

Next, calculate lbs/day solids leaving the thickener:

<u>Via the underflow</u>

$$(45 \text{ gpm})(1440 \text{ min/day})(8.34 \text{ lbs/gal})\left(\frac{7.4}{100}\right) = 39{,}992 \text{ lbs/day}$$

<u>Via the effluent flow</u>

$$(105 \text{ gpm} - 45 \text{ gpm})(1440 \text{ min/day})(8.34 \text{ lbs/gal})\left(\frac{0.068}{100}\right) = 490 \text{ lbs/day}$$

To summarize:

Entering thickener = 37,830 lbs/day

Leaving thickener = 40,482 lbs/day
(39,992 lbs + 490 lbs)

Since more solids are leaving the thickener than entering, the sludge blanket will <u>drop</u>.

b) The lbs/day represented by the drop are:

$$\underset{\text{Leaving}}{40{,}482 \text{ lbs/day}} - \underset{\text{Entering}}{37{,}830 \text{ lbs/day}} = \underset{\text{Drop}}{2652 \text{ lbs/day}}$$

14.
$$\frac{(50 \text{ gpm})(1440 \frac{\text{min}}{\text{day}})(8.34 \frac{\text{lbs}}{\text{gal}})(\frac{3.8}{100})}{(0.785)(30 \text{ ft})(30 \text{ ft})} = 32.3 \frac{\text{lbs/day}}{\text{sq ft}}$$

15.
$$\frac{180{,}000 \text{ gpd}}{1440 \text{ min/day}} = 125 \text{ gpm}$$

Now hydraulic loading rate in gpm/sq ft can be calculated:

$$\frac{125 \text{ gpm}}{(50 \text{ ft})(15 \text{ ft})} = 0.2 \text{ gpm/sq ft}$$

16.
$$\frac{(0.785)(25 \text{ ft})(25 \text{ ft})(2.4 \text{ ft})(7.48 \frac{\text{gal}}{\text{cu ft}})(8.34 \frac{\text{lbs}}{\text{gal}})(\frac{6.7}{100})}{\frac{9200 \text{ lbs/day}}{24 \text{ hrs/day}}} = 12.8 \text{ hrs}$$

17.
$$\frac{80{,}000 \text{ gpd}}{24 \text{ hrs/day}} = 3333 \text{ gph}$$

18.
$$\frac{(5 \text{ cfm})(0.075 \text{ lbs/cu ft})}{(100 \text{ gpm})(8.34 \text{ lbs/gal})(\frac{0.72}{100})} = 0.06$$

CHAPTER 15—ACHIEVEMENT TEST—Cont'd

19. $110 = \dfrac{x \text{ gpm}}{70 \text{ gpm}} \times 100$

$\dfrac{(110)(70)}{100} = x$

$77 \text{ gpm} = x$

20. $\dfrac{(77{,}000 \text{ gpd})}{24 \text{ hrs/day}} \dfrac{(28 \text{ min})}{30 \text{ min}} = 2994 \text{ gal/hr}$

21. First calculate desired stored solids:

$\dfrac{2.3 \text{ ft}}{5 \text{ ft}} = \dfrac{x \text{ lbs/min stored solids}}{45 \text{ lbs/min solids entering}}$

$\dfrac{(2.3)(45)}{5} = x$

$21 \text{ lbs/min} = x$
Storage Rate

Next, calculate the solids withdrawal rate required to achieve a storage rate of 21 lbs/min:

$45 \text{ lbs/min} = x \text{ lbs/min} + 21 \text{ lbs/min}$

$24 \text{ lbs/min} = x$
Solids Withd.

Now the gpm sludge withdrawal rate can be determined:

$\underset{\substack{\text{Sludge} \\ \text{Withd.}}}{(x \text{ gpm})} \underset{\text{lbs/gal}}{(8.34)} \dfrac{(7)}{100} = \underset{\text{Solids Withd.}}{24 \text{ lbs/min}}$

$x = 41 \text{ gpm Sludge Withdrawal rate}$

22. $\dfrac{(108{,}000 \text{ gpd})(8.34 \text{ lbs})(0.695)}{24 \text{ hrs/day} \quad \text{gal} \quad 100} = 261 \text{ lbs/hr}$

23. $\dfrac{(30 \text{ cu ft})(7.48 \text{ gal})(8.34 \text{ lbs})(6.4)}{\dfrac{\text{cu ft} \qquad \text{gal} \quad 100}{(60 \text{ gpm})(8.34 \text{ lbs})(0.71)}} = 34 \text{ min}$
$\qquad\qquad\qquad\qquad \text{gal} \quad 100$

24. $\dfrac{(90 \text{ gpm})(60 \text{ min})(8.34 \text{ lbs})(0.77)(23 \text{ min})}{\text{hr} \qquad \text{gal} \quad 100 \quad 24.5 \text{ min}} = 326 \text{ lbs/hr}$

25. $\dfrac{(14 \text{ cu ft}) (62.4) \dfrac{(3.6)}{100} + (4 \text{ cu ft}) (62.4) \dfrac{(7.6)}{100}}{\substack{\text{lbs/cu ft} \qquad\qquad \text{lbs/cu ft}}}$
$\qquad \dfrac{}{(14 \text{ cu ft})(62.4 \text{ lbs}) + (4 \text{ cu ft})(62.4 \text{ lbs})} \times 100$
$\qquad\qquad\qquad \text{cu ft} \qquad\qquad\quad \text{cu ft}$

$$= \frac{31.4 \text{ lbs} + 19.0 \text{ lbs}}{873.6 \text{ lbs} + 249.6 \text{ lbs}} \times 100$$

$$= 4.5\%$$

Chapter 16

PRACTICE PROBLEMS 16.1

1. $$\frac{(4120 \text{ gpd})(8.34 \frac{\text{lbs}}{\text{gal}})(\frac{5.6}{100}) + (6740 \text{ gpd})(8.34 \frac{\text{lbs}}{\text{gal}})(\frac{3.4}{100})}{(4120 \text{ gpd})(8.34 \frac{\text{lbs}}{\text{gal}}) + (6740 \text{ gpd})(8.34 \frac{\text{lbs}}{\text{gal}})} \times 100$$

$$= \frac{1924 \text{ lbs/day} + 1911 \text{ lbs/day}}{34,360 \text{ lbs/day} + 56,212 \text{ lbs/day}} \cdot \times 100$$

$$= 4.2\%$$

2. $$\frac{(3430 \text{ gpd})(8.34 \frac{\text{lbs}}{\text{gal}})(\frac{5.4}{100}) + (5190 \text{ gpd})(8.34 \frac{\text{lbs}}{\text{gal}})(\frac{3.9}{100})}{(3430 \text{ gpd})(8.34 \frac{\text{lbs}}{\text{gal}}) + (5790 \text{ gpd})(8.34 \frac{\text{lbs}}{\text{gal}})} \times 100$$

$$= \frac{1545 \text{ lbs/day} + 1688 \text{ lbs/day}}{28,606 \text{ lbs/day} + 43,285 \text{ lbs/day}} \times 100$$

$$= 4.5\%$$

3. $$\frac{(3840 \text{ gpd})(8.34 \frac{\text{lbs}}{\text{gal}})(\frac{6.1}{100}) + (6670 \text{ gpd})(8.34 \frac{\text{lbs}}{\text{gal}})(\frac{4.6}{100})}{(3840 \text{ gpd})(8.34 \frac{\text{lbs}}{\text{gal}}) + (6670 \text{ gpd})(8.34 \frac{\text{lbs}}{\text{gal}})} \times 100$$

$$= \frac{1954 \text{ lbs/day} + 2559 \text{ lbs/day}}{32,026 \text{ lbs/day} + 55,628 \text{ lbs/day}} \times 100$$

$$= 5.1\%$$

4. $$\frac{(2470 \text{ gpd})(8.34 \frac{\text{lbs}}{\text{gal}})(\frac{4.2}{100}) + (3510 \text{ gpd})(8.62 \frac{\text{lbs}}{\text{gal}})(\frac{5.8}{100})}{(2470 \text{ gpd})(8.34 \frac{\text{lbs}}{\text{gal}}) + (3510 \text{ gpd})(8.62 \frac{\text{lbs}}{\text{gal}})} \times 100$$

$$= \frac{865 \text{ lbs/day} + 1755 \text{ lbs/day}}{20,600 \text{ lbs/day} + 30,256 \text{ lbs/day}} \times 100$$

$$= 5.2\%$$

PRACTICE PROBLEMS 16.2

1. $\dfrac{(0.8 \text{ gal})}{\text{stroke}} \dfrac{(30 \text{ strokes})}{\text{min}} = 24 \text{ gpm}$

2. $(0.785)(0.83 \text{ ft})(0.83 \text{ ft})(0.25 \text{ ft})(7.48 \text{ gal/cu ft}) = 1 \text{ gal/stroke}$

   Then: $\dfrac{(1 \text{ gal})}{\text{stroke}} \dfrac{(32 \text{ strokes})}{\text{min}} = 32 \text{ gpm}$

3. $(0.785)(0.67 \text{ ft})(0.67 \text{ ft})(0.33 \text{ ft})(7.48 \text{ gal/cu ft}) = 0.9 \text{ gal/stroke}$

   Then: $\dfrac{(0.9 \text{ gal})}{\text{stroke}} \dfrac{(45 \text{ strokes})}{\text{min}} \dfrac{(120 \text{ min})}{\text{day}} = 4860 \text{ gpd}$

4. $(0.785)(1 \text{ ft})(1 \text{ ft})(0.42 \text{ ft})(7.48 \text{ gal/cu ft}) = 2.5 \text{ gal/stroke}$

   And: $\dfrac{(2.5 \text{ gal})}{\text{stroke}} \dfrac{(35 \text{ strokes})}{\text{min}} \dfrac{(150 \text{ min})}{\text{day}} = 13,125 \text{ gpd}$

PRACTICE PROBLEMS 16.3

1. First calculate min/day pumping rate required:

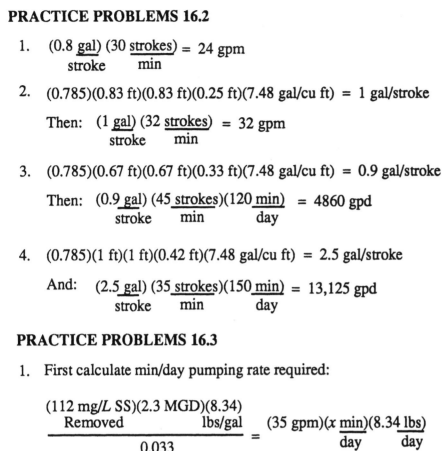

$$\dfrac{(112 \text{ mg/L SS})(2.3 \text{ MGD})(8.34 \text{ lbs/gal})}{0.033} = \dfrac{(35 \text{ gpm})(x \text{ min})}{\text{day}}\dfrac{(8.34 \text{ lbs})}{\text{day}}$$

$$\dfrac{(112)(2.3)(8.34)}{(0.033)(35)(8.34)} = x$$

$$223 \dfrac{\text{min}}{\text{day}} = x$$

Then convert min/day pumping rate to min/hr pumping rate:

$$\dfrac{223 \text{ min/day}}{24 \text{ hrs/day}} = 9 \text{ min/hr}$$

2. First calculate min/day pumping rate required:

$$\dfrac{(110 \text{ mg/L SS})(1.95 \text{ MGD})(8.34 \text{ lbs/gal})}{0.038} = \dfrac{(30 \text{ gpm})(x \text{ min})}{\text{day}}\dfrac{(8.34 \text{ lbs})}{\text{day}}$$

$$\dfrac{(110)(1.95)(8.34)}{(0.038)(30)(8.34)} = x$$

$$188 \dfrac{\text{min}}{\text{day}} = x$$

Then convert min/day pumping rate to min/hr pumping rate:

$$\dfrac{188 \text{ min/day}}{24 \text{ hrs/day}} = 8 \text{ min/hr}$$

3. First calculate min/day pumping rate required:

$$\frac{\underset{\text{Removed}}{(124 \text{ mg/}L)}(3.45 \text{ MGD})\underset{\text{lbs/gal}}{(8.34)}}{\dfrac{4.1}{100}} = (40 \text{ gpm})(x \underset{\text{day}}{\text{min}})(8.34 \underset{\text{day}}{\text{lbs}})$$

$$\frac{(124)(3.45)(8.34)}{(0.041)(40)(8.34)} = x$$

$$261 \underset{\text{day}}{\text{min}} = x$$

Then convert min/day pumping rate to min/hr:

$$\frac{261 \text{ min/day}}{24 \text{ hrs/day}} = 11 \text{ min/hr}$$

4. First calculate min/day pumping rate required:

$$\frac{\underset{\text{Removed}}{(123 \text{ mg/}L)}(1.3 \text{ MGD})\underset{\text{lbs/gal}}{(8.34)}}{\dfrac{3.5}{100}} = (35 \text{ gpm})(x \underset{\text{day}}{\text{min}})(8.34 \underset{\text{day}}{\text{lbs}})$$

$$\frac{(123)(1.3)(8.34)}{(0.035)(35)(8.34)} = 131 \text{ min/day}$$

Then convert min/day pumping rate required:

$$\frac{131 \text{ min/day}}{24 \text{ hrs/day}} = 5.5 \text{ min/hr}$$

PRACTICE PROBLEMS 16.4

1. $(8450 \text{ lbs/day Solids})\dfrac{(67)}{100} = 5662 \text{ lbs/day VS}$

2. $(2780 \text{ lbs/day Solids})\dfrac{(65)}{100} = 1807 \text{ lbs/day VS}$

3. $(3630 \text{ gpd Sludge})(8.34 \underset{\text{gal}}{\text{lbs}}) \dfrac{(5.7)}{100} \dfrac{(71)}{100} = 1225 \text{ lbs/day VS}$

4. $(5,015 \text{ gpd Sludge})(8.34 \underset{\text{gal}}{\text{lbs}}) \dfrac{(6)}{100} \dfrac{(69)}{100} = 1732 \text{ lbs/day VS}$

PRACTICE PROBLEMS 16.5

1. $25 = \dfrac{x \text{ gal}}{293,590 \text{ gal}} \times 100$

$$x = \frac{(293,590 \text{ gal})(25)}{100}$$

$$x = 73,398 \text{ gal}$$

PRACTICE PROBLEMS 16.5—Cont'd

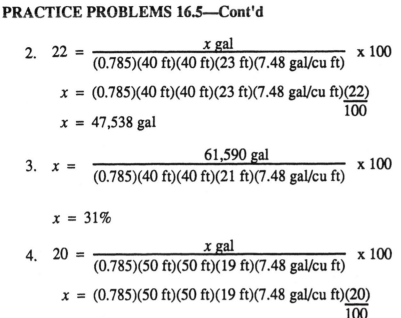

2. $22 = \dfrac{x \text{ gal}}{(0.785)(40 \text{ ft})(40 \text{ ft})(23 \text{ ft})(7.48 \text{ gal/cu ft})} \times 100$

$x = \dfrac{(0.785)(40 \text{ ft})(40 \text{ ft})(23 \text{ ft})(7.48 \text{ gal/cu ft})(22)}{100}$

$x = 47{,}538 \text{ gal}$

3. $x = \dfrac{61{,}590 \text{ gal}}{(0.785)(40 \text{ ft})(40 \text{ ft})(21 \text{ ft})(7.48 \text{ gal/cu ft})} \times 100$

$x = 31\%$

4. $20 = \dfrac{x \text{ gal}}{(0.785)(50 \text{ ft})(50 \text{ ft})(19 \text{ ft})(7.48 \text{ gal/cu ft})} \times 100$

$x = \dfrac{(0.785)(50 \text{ ft})(50 \text{ ft})(19 \text{ ft})(7.48 \text{ gal/cu ft})(20)}{100}$

$x = 55{,}782 \text{ gal}$

PRACTICE PROBLEMS 16.6

1. $\dfrac{(65{,}570 \text{ lbs/day Sludge})\dfrac{(5.2)}{100}\dfrac{(69)}{100}}{(100{,}000 \text{ gal})(8.34 \dfrac{\text{lbs}}{\text{gal}})\dfrac{(6.2)}{100}\dfrac{(55)}{100}}$

$= \dfrac{2353 \text{ lbs VS/day}}{28{,}439 \text{ lbs VS}}$

$= 0.08 \dfrac{\text{lbs VS Added/day}}{\text{lb VS in Digester}}$

2. $\dfrac{0.05 \text{ lbs/day}}{\text{lb VS}} = \dfrac{x \text{ lbs/day VS}}{(21{,}190 \text{ gal})(8.34 \dfrac{\text{lbs}}{\text{gal}})\dfrac{(6.1)}{100}\dfrac{(56)}{100}}$

$(0.05)(21{,}190)(8.34 \text{ lbs/gal})\dfrac{(6.1)}{100}\dfrac{(56)}{100} = x \text{ lbs/day}$

$302 \text{ lbs VS/day} = x$

3. $\dfrac{(59{,}880 \text{ lbs/day Sludge})(5.2)(68)}{\dfrac{100 \ 100}{(95{,}000 \text{ gal})(8.34 \ \frac{\text{lbs}}{\text{gal}})(\frac{6}{100})(\frac{59}{100})}}$

$= \dfrac{2117 \text{ lbs VS/day}}{28{,}047 \text{ lbs VS}}$

$= 0.08 \ \dfrac{\text{lbs VS Added/day}}{\text{lb VS in Digester}}$

4. $\dfrac{0.09 \text{ lbs VS Added/day}}{\text{lb VS in Digester}} = \dfrac{(910 \text{ gpd})(8.34 \text{ lbs/gal})(5.6)(70)}{\dfrac{100 \ 100}{(x \text{ gal})(8.34 \text{ lbs/gal})(8.5)(54)}}$

$x = \dfrac{(910)(8.34)(0.056)(0.70)}{(0.09)(8.85)(0.085)(0.54)}$

$x = 8138 \text{ gal seed sludge}$

PRACTICE PROBLEMS 16.7

1. $\dfrac{(85{,}460 \text{ lbs/day Sludge})(\frac{6}{100})(\frac{71}{100})}{(0.785)(40 \text{ ft})(40 \text{ ft})(21 \text{ ft})}$

$= \dfrac{0.14 \text{ lbs VS/day}}{\text{cu ft}}$

2. $\dfrac{(38{,}120 \text{ gpd})(8.34 \text{ lbs/gal})(5.7)(69)}{\dfrac{100 \ 100}{(0.785)(50 \text{ ft})(50 \text{ ft})(22 \text{ ft})}}$

$= \dfrac{0.290 \text{ lbs VS/day}}{\text{cu ft}}$

Then convert to lbs VS/day/1000 cu ft

$\dfrac{(0.290 \text{ lbs VS/day})}{1 \text{ cu ft}} \ \dfrac{\text{x } 1000}{\text{x } 1000} = \dfrac{290 \text{ lbs/day VS}}{1000 \text{ cu ft}}$

PRACTICE PROBLEMS 16.7—Cont'd

3.
$$\frac{(29{,}900 \text{ gpd})(8.34 \text{ lbs/gal})\dfrac{(5.8)}{100}\dfrac{(70)}{100}}{(0.785)(45 \text{ ft})(45 \text{ ft})(20 \text{ ft})}$$

$$= \frac{0.318 \text{ lbs VS/day}}{\text{cu ft}}$$

Then convert to lbs VS/day/1000 cu ft:

$$\frac{(0.318 \text{ lbs VS/day})}{1 \text{ cu ft}} \quad \frac{\times 1000}{\times 1000} \quad = \quad \frac{318 \text{ lbs VS/day}}{1000 \text{ cu ft}}$$

4.
$$\frac{(17{,}000 \text{ gpd Sludge})(8.34 \text{ lbs/gal})\dfrac{(5.2)}{100}\dfrac{(70)}{100}}{(0.785)(40 \text{ ft})(40 \text{ ft})(19 \text{ ft})}$$

$$= \frac{0.216 \text{ lbs/day VS}}{\text{cu ft}}$$

Then convert to lbs/day VS/1000 cu ft:

$$\frac{(0.216 \text{ lbs/day VS})}{1 \text{ cu ft}} \quad \frac{\times 1000}{\times 1000} \quad = \quad \frac{216 \text{ lbs/day VS}}{1000 \text{ cu ft}}$$

PRACTICE PROBLEMS 16.8

1.
$$\frac{1 \text{ lb/day VS}}{10 \text{ lbs Dig. Sludge}} = \frac{(2800 \text{ gpd})(8.34 \text{ lbs/gal})\dfrac{(5.6)}{100}\dfrac{(68)}{100}}{x \text{ lbs Digested Sludge}}$$

$$x = (2800)(8.34)(0.056)(0.68)(10)$$

$$x = 8892 \text{ lbs Digested Sludge}$$

2.
$$\frac{1 \text{ lb/day VS}}{10 \text{ lbs Dig. Sludge}} = \frac{(6200 \text{ gpd})(8.34 \text{ lbs/gal})\dfrac{(6)}{100}\dfrac{(72)}{100}}{x \text{ lbs Digested Sludge}}$$

$$x = (6200)(8.34)(0.06)(0.72)(10)$$

$$x = 22{,}338 \text{ lbs Digested Sludge}$$

3. $\dfrac{1\ \text{lb/day VS}}{10\ \text{lbs Dig. Sludge}} = \dfrac{(5100\ \text{gpd})(8.34\ \text{lbs/gal})\dfrac{(6.3)}{100}\dfrac{(69)}{100}}{x\ \text{lbs Digested Sludge}}$

$x = (5100)(8.34)(0.063)(0.69)(10)$

$x = 18{,}490\ \text{lbs Digested Sludge}$

4. $\dfrac{1\ \text{lb/day VS}}{10\ \text{lbs Dig. Sludge}} = \dfrac{(3900\ \text{gpd})(8.34\ \text{lbs/gal})\dfrac{(7)}{100}\dfrac{(70)}{100}}{x\ \text{lbs Digested Sludge}}$

$x = (3900)(8.34)(0.07)(0.70)(10)$

$x = 15{,}938\ \text{lbs Digested Sludge}$

PRACTICE PROBLEMS 16.9

1. $\dfrac{172\ \text{mg/}L}{2130\ \text{mg/}L} = 0.08$

2. $\dfrac{150\ \text{mg/}L}{2490\ \text{mg/}L} = 0.06$

3. $\dfrac{146\ \text{mg/}L}{2320\ \text{mg/}L} = 0.06$

4. $\dfrac{181\ \text{mg/}L}{2560\ \text{mg/}L} = 0.07$

PRACTICE PROBLEMS 16.10

1. $(2270\ \text{mg/}L)(0.242\ \text{MG})(8.34\ \text{lbs/gal}) = 4581\ \text{lbs Lime}$

2. $(1990\ \text{mg/}L)(0.195\ \text{MG})(8.34\ \text{lbs/gal}) = 3236\ \text{lbs Lime}$

3. $(2510\ \text{mg/}L)(0.225\ \text{MG})(8.34\ \text{lbs/gal}) = 4710\ \text{lbs Lime}$

4. $(2360\ \text{mg/}L)(0.18\ \text{MG})(8.34\ \text{lbs/gal}) = 3543\ \text{lbs Lime}$

PRACTICE PROBLEMS 16.11

1. $\dfrac{0.66 - 0.51}{0.66 - (0.66)(0.51)} \times 100 = \dfrac{0.15}{0.3234} \times 100 = 46\%\ \text{VS Reduction}$

PRACTICE PROBLEMS—Cont'd

2. $\dfrac{0.72 - 0.53}{0.72 - (0.72)(0.53)} \times 100 = \dfrac{0.19}{0.3384} \times 100 = 56\%$ VS Reduction

3. $\dfrac{0.69 - 0.53}{0.69 - (0.69)(0.53)} \times 100 = \dfrac{0.16}{0.3243} \times 100 = 49\%$ VS Reduction

4. $\dfrac{0.67 - 0.52}{0.67 - (0.67)(0.52)} \times 100 = \dfrac{0.15}{0.3216} \times 100 = 47\%$ VS Reduction

PRACTICE PROBLEMS 16.12

1. $\dfrac{(3700 \text{ gpd})(8.34 \text{ lbs/gal})\dfrac{(6.1)}{100}\dfrac{(71)}{100}\dfrac{(55)}{100}}{35,000 \text{ cu ft}} = 0.02 \dfrac{\text{lbs VS Destroyed/day}}{\text{cu ft}}$

2. $\dfrac{(4450 \text{ gpd})(8.34 \text{ lbs/gal})\dfrac{(6)}{100}\dfrac{(68)}{100}\dfrac{(56)}{100}}{33,000 \text{ cu ft}} = 0.03 \dfrac{\text{lbs VS Destroyed/day}}{\text{cu ft}}$

3. $\dfrac{(2800 \text{ gpd})(8.34 \text{ lbs/gal})\dfrac{(5.8)}{100}\dfrac{(70)}{100}\dfrac{(54)}{100}}{(0.785)(50 \text{ ft})(50 \text{ ft})(20 \text{ ft})} = 0.01 \dfrac{\text{lbs VS Destroyed/day}}{\text{cu ft}}$

4. First calculate lbs VS destroyed/day/cu ft:

$\dfrac{(3000 \text{ gpd})(8.34 \text{ lbs/gal})\dfrac{(6.3)}{100}\dfrac{(67)}{100}\dfrac{(58)}{100}}{(0.785)(45 \text{ ft})(45 \text{ ft})(18 \text{ ft})} = 0.021 \dfrac{\text{lbs VS Destroyed/day}}{\text{cu ft}}$

Then convert to lbs VS destroyed/day/1000 cu ft:

$\dfrac{(0.021 \text{ lbs VS Destroyed/day})}{1 \text{ cu ft}} \times \dfrac{1000}{1000} = \dfrac{21 \text{ lbs VS Destroyed/day}}{1000 \text{ cu ft}}$

PRACTICE PROBLEMS 16.13

1. $\dfrac{6620 \text{ cu ft gas/day}}{520 \text{ lbs VS Destroyed/day}} = \dfrac{12.7 \text{ cu ft}}{\text{lbs VS Destroyed}}$

2. $\dfrac{19,150 \text{ cu ft gas/day}}{(2060 \text{ lbs VS/day})\dfrac{(57)}{100}} = \dfrac{16.3 \text{ cu ft}}{\text{lbs VS Destroyed}}$

3. $\dfrac{8610 \text{ cu ft gas/day}}{574 \text{ lbs VS Destroyed/day}} = \dfrac{15 \text{ cu ft}}{\text{lbs VS Destroyed}}$

4. $\dfrac{25,640 \text{ cu ft gas/day}}{\dfrac{(3270 \text{ lbs VS/day})(56)}{100}} = \dfrac{14 \text{ cu ft}}{\text{lbs VS Destroyed}}$

PRACTICE PROBLEMS 16.14

1. First calculate solids and water entering the digester:

   • lbs/day total solids entering:

   $(27,100 \text{ lbs/day})\underset{\text{Sludge}}{\dfrac{(5.8)}{100}} = 1572$ lbs/day Total Solids

   • lbs/day volatile solids entering:

   $(1572 \text{ lbs/day})\underset{\text{Solids}}{\dfrac{(68)}{100}} = 1069$ lbs/day Volatile Solids

   • lbs/day fixed solids entering:

   $\underset{\text{Tot. Solids}}{1572 \text{ lbs/day}} - \underset{\text{Vol. Solids}}{1069 \text{ lbs/day}} = 503$ lbs/day Fixed Solids

   • lbs/day water entering:

   $\underset{\text{Sludge}}{27,100 \text{ lbs/day}} - \underset{\text{Tot. Solids}}{1572 \text{ lbs/day}} = 25,528$ lbs/day Water

   Then calculate solids, water and gas leaving the digester:
   (To begin these calculations, you must first determine the percent volatile solids reduction during digestion.)

   • *% Volatile Solids Reduction*

   $$\underset{\text{Reduction}}{\% \text{ VS}} = \frac{\text{In} - \text{Out}}{\text{In} - (\text{In x Out})} \text{ x } 100$$

   $$= \frac{0.68 - 0.56}{0.68 - (0.68)(0.56)} \text{ x } 100$$

   $$= \frac{0.12}{0.2992} \text{ x } 100$$

   $$= 40\% \text{ VS Reduction}$$

   • *lbs Gas Produced*
   (VS destroyed or reduced = lbs gas produced)

   $$\frac{(\text{lbs VS Entering Digester}) (\% \text{ VS Reduction})}{100} = \text{lbs Gas Produced}$$

   $$\frac{(1069 \text{ lbs Vol. Sol.}) (40)}{100} = \begin{array}{l} 428 \text{ lbs} \\ \text{Gas Produced} \end{array}$$

(Answer Cont'd On Next Page)

PRACTICE PROBLEMS 16.14—Cont'd

- *lbs Volatile Solids in Digested Sludge*

  lbs VS Entering Digester – lbs VS Destroyed = lbs VS Leaving Digester

  1069 lbs – 428 lbs = 641 lbs VS in Digested Sludge

- *lbs Total Solids in Digested Sludge*

$$\frac{\text{lbs Vol. Sol.}}{\dfrac{\% \text{ VS}}{100}} = \text{lbs Total Solids}$$

$$\frac{641 \text{ lbs VS}}{\dfrac{56}{100}} = 1145 \text{ lbs Total Solids}$$

- *lbs Fixed Solids in Digested Sludge*

  Total Solids, lbs – Volatile Solids, lbs = Fixed Solids, lbs

  1145 lbs – 641 lbs = 504 lbs

- *lbs Digested Sludge*

$$\frac{\text{lbs Total Solids}}{\dfrac{\% \text{ Solids}}{100}} = \text{lbs Digested Sludge}$$

$$\frac{1145 \text{ lbs Tot. Sol.}}{\dfrac{4.5}{100}} = 25,444 \text{ lbs Digested Sludge}$$

- *lbs Water in Digested Sludge*

  Sludge, lbs – Total Solids, lbs = Water, lbs

  25,444 lbs – 1145 lbs = 24,299 lbs

Summary Comparison	
Sludge Entering	Sludge Leaving
Tot. Sol.—1572 lbs (VS—1069 lbs) (FS— 503 lbs) Water—25,528 lbs $\quad$ 27,100 lbs	Tot. Sol.—1145 lbs (VS—641 lbs) (FS— 504 lbs) Water—24,299 lbs Gas— 428 lbs $\quad$ 25,872 lbs

2. <u>First calculate solids and water entering the digester:</u>

• lbs/day total solids entering:

$$(28,000 \text{ lbs/day})\underset{\text{Sludge}}{\left(\frac{6}{100}\right)} = 1680 \text{ lbs/day Total Solids}$$

• lbs/day volatile solids entering:

$$(1680 \text{ lbs/day})\underset{\text{Solids}}{\left(\frac{71}{100}\right)} = 1193 \text{ lbs/day Volatile Solids}$$

• lbs/day fixed solids entering:

$$\underset{\text{Tot. Solids}}{1680 \text{ lbs/day}} - \underset{\text{Vol. Solids}}{1193 \text{ lbs/day}} = 487 \text{ lbs/day Fixed Solids}$$

• lbs/day water entering:

$$\underset{\text{Sludge}}{28,000 \text{ lbs/day}} - \underset{\text{Tot. Solids}}{1680 \text{ lbs/day}} = 26,320 \text{ lbs/day Water}$$

<u>Then calculate solids, water and gas leaving the digester:</u>
(To begin these calculations, you must first determine the percent volatile solids reduction during digestion.)

• *% Volatile Solids Reduction*

$$\frac{\% \text{ VS}}{\text{Reduction}} = \frac{\text{In} - \text{Out}}{\text{In} - (\text{In} \times \text{Out})} \times 100$$

$$= \frac{0.71 - 0.58}{0.71 - (0.71)(0.58)} \times 100$$

$$= \frac{0.13}{0.2982} \times 100$$

$$= 44\% \text{ VS Reduction}$$

• *lbs Gas Produced*
(VS destroyed or reduced = lbs gas produced)

$$(\text{lbs VS Entering Digester}) \frac{(\% \text{ VS Reduction})}{100} = \text{lbs Gas Produced}$$

$$(1193 \text{ lbs Vol. Sol.}) \frac{(44)}{100} = 525 \text{ lbs Gas Produced}$$

• *lbs Volatile Solids in Digested Sludge*

$$\text{lbs VS Entering Digester} - \text{lbs VS Destroyed} = \text{lbs VS Leaving Digester}$$

$$1193 \text{ lbs} - 525 \text{ lbs} = 668 \text{ lbs VS in Digested Sludge}$$

(Answer Cont'd On Next Page)

PRACTICE PROBLEMS 16.14—Cont'd

- *lbs Total Solids in Digested Sludge*

$$\frac{\text{lbs Vol. Sol.}}{\dfrac{\% \text{ VS}}{100}} = \text{lbs Total Solids}$$

$$\frac{668 \text{ lbs VS}}{\dfrac{58}{100}} = 1152 \text{ lbs Total Solids}$$

- *lbs Fixed Solids in Digested Sludge*

Total Solids, lbs – Volatile Solids, lbs = Fixed Solids, lbs

1152 lbs – 668 lbs = 484 lbs

- *lbs Digested Sludge*

$$\frac{\text{lbs Total Solids}}{\dfrac{\% \text{ Solids}}{100}} = \text{lbs Digested Sludge}$$

$$\frac{1152 \text{ lbs Tot. Sol.}}{\dfrac{4.3}{100}} = 26{,}791 \text{ lbs Digested Sludge}$$

- *lbs Water in Digested Sludge*

Sludge, lbs – Total Solids, lbs = Water, lbs

26,791 lbs – 1152 lbs = 25,639 lbs

Summary Comparison	
Sludge Entering	Sludge Leaving

Tot. Sol.—1680 lbs
(VS—1193 lbs)
(FS— 487 lbs)
 	
Water—26,320 lbs
        28,000 lbs
	Tot. Sol.—1152 lbs
	(VS—668 lbs)
	(FS— 484 lbs)
	
	Water—25,639 lbs
	Gas— 525 lbs
	        27,316 lbs

PRACTICE PROBLEMS 16.15

1. $\dfrac{(0.785)(50\text{ ft})(50\text{ ft})(10\text{ ft})(7.48\text{ gal/cu ft})}{8900\text{ gpd}} = 16.5\text{ days}$

2. $\dfrac{(0.785)(40\text{ ft})(40\text{ ft})(12\text{ ft})(7.48\text{ gal/cu ft})}{8100\text{ gpd}} = 13.9\text{ days}$

3. $\dfrac{(90\text{ ft})(20\text{ ft})(10\text{ ft})(7.48\text{ gal/cu ft})}{7600\text{ gpd}} = 17.7\text{ days}$

4. Digestion time for sludge with 3.2% solids:

$$\dfrac{(0.785)(35\text{ ft})(35\text{ ft})(10\text{ ft})(7.48\text{ gal/cu ft})}{10{,}000\text{ gpd}} = 7.2\text{ days}$$

Digestion time for sludge with 6% solids:

$$\dfrac{(0.785)(35\text{ ft})(35\text{ ft})(10\text{ ft})(7.48\text{ gal/cu ft})}{5300\text{ gpd}} = 13.6\text{ days}$$

Thus the sludge with the higher solids content permits greater time for digestion.

PRACTICE PROBLEMS 16.16

1. $\dfrac{0.04\text{ cfm}}{1\text{ cu ft Dig. Vol.}} = \dfrac{x\text{ cfm Air Required}}{(85\text{ ft})(25\text{ ft})(10\text{ ft})}$

$(0.04)(85)(25)(10) = x$

$\begin{array}{c}850\text{ cfm} = x \\ \text{Air}\end{array}$

2. $\dfrac{45\text{ cfm}}{1000\text{ cu ft}} = \dfrac{x\text{ cfm Air Required}}{(0.785)(60\text{ ft})(60\text{ ft})(10\text{ ft})}$

$\dfrac{(45)(0.785)(60\text{ ft})(60\text{ ft})(10\text{ ft})}{1000} = x$

$\begin{array}{c}1272\text{ cfm} = x \\ \text{Air}\end{array}$

3. $\begin{array}{l} O_2\text{ Uptake} \\ \text{mg/L/hr} \end{array} = \dfrac{5.2\text{ mg/L} - 3.1\text{ mg/L}}{3\text{ min}} \text{ x } \dfrac{60\text{ min}}{\text{hr}}$

$\qquad\qquad = \dfrac{(2.1)(60)}{3}$

$\qquad\qquad = 42\text{ mg/L/hr}$

PRACTICE PROBLEMS 16.16—Cont'd

4. O_2 Uptake $= \dfrac{5.3\ mg/L - 3.5\ mg/L}{3\ min} \times \dfrac{60\ min}{hr}$
    mg/L/hr

$\qquad = 36\ mg/L/hr$

PRACTICE PROBLEMS 16.17

1. $(25\ mg/L)(0.11\ MG)(8.34\ lbs/gal) = 22.9$ lbs Caustic

2. $(17\ mg/L)(0.147\ MG)(8.34\ lbs/gal) = 20.8$ lbs Caustic

3. $\dfrac{(62\ mg)}{2\ L}(0.053\ MG)(8.34\ lbs/gal) = 13.7$ lbs Caustic

4. First calculate the gallon volume of the digester:

$(0.785)(50\ ft)(50\ ft)(12\ ft)(7.48\ gal/cu\ ft) = 176,154$ gal

Then calculate lbs caustic required:

$\dfrac{(80\ mg)}{2\ L}(0.176\ MG)(8.34\ lbs/gal) = 58.7$ lbs Caustic

CHAPTER 16—ACHIEVEMENT TEST

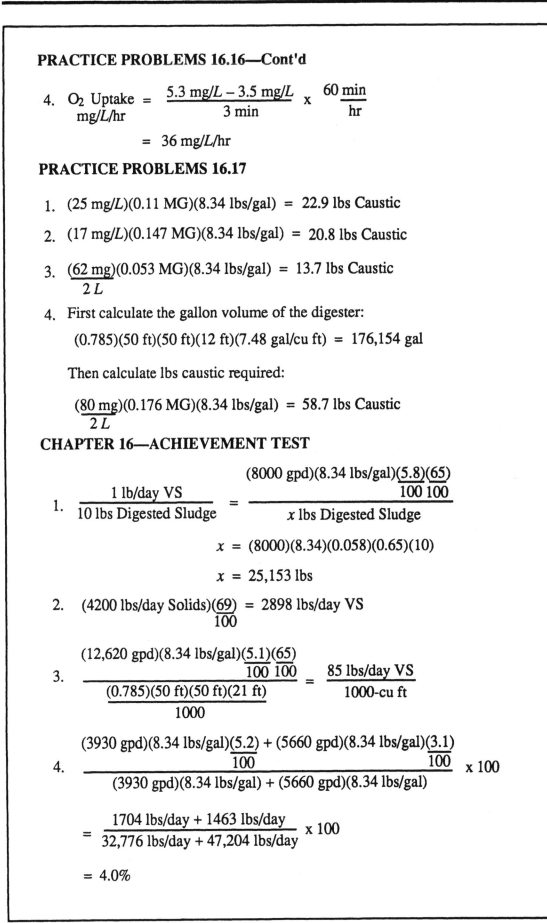

1. $\dfrac{1\ lb/day\ VS}{10\ lbs\ Digested\ Sludge} = \dfrac{(8000\ gpd)(8.34\ lbs/gal)\dfrac{(5.8)}{100}\dfrac{(65)}{100}}{x\ lbs\ Digested\ Sludge}$

$\qquad x = (8000)(8.34)(0.058)(0.65)(10)$

$\qquad x = 25,153$ lbs

2. $(4200\ lbs/day\ Solids)\dfrac{(69)}{100} = 2898$ lbs/day VS

3. $\dfrac{\dfrac{(12,620\ gpd)(8.34\ lbs/gal)\dfrac{(5.1)}{100}\dfrac{(65)}{100}}{\dfrac{(0.785)(50\ ft)(50\ ft)(21\ ft)}{1000}}}{} = \dfrac{85\ lbs/day\ VS}{1000\text{-}cu\ ft}$

4. $\dfrac{(3930\ gpd)(8.34\ lbs/gal)\dfrac{(5.2)}{100} + (5660\ gpd)(8.34\ lbs/gal)\dfrac{(3.1)}{100}}{(3930\ gpd)(8.34\ lbs/gal) + (5660\ gpd)(8.34\ lbs/gal)} \times 100$

$\qquad = \dfrac{1704\ lbs/day + 1463\ lbs/day}{32,776\ lbs/day + 47,204\ lbs/day} \times 100$

$\qquad = 4.0\%$

5. $(0.785)(0.67 \text{ ft})(0.67 \text{ ft})(0.42 \text{ ft})(7.48 \text{ gal/cu ft}) = 1.1 \text{ gal/stroke}$

    Then:

    $(1.1 \dfrac{\text{gal}}{\text{stroke}})(3400 \dfrac{\text{strokes}}{\text{day}}) = 3740 \text{ gpd}$

6. $x = \dfrac{87,200 \text{ gal}}{(0.785)(50 \text{ ft})(50 \text{ ft})(22 \text{ ft})(7.48 \text{ gal/cu ft})} \times 100$

    $x = 27\%$

7. $\dfrac{(3600 \text{ gpd sludge})(8.34 \text{ lbs/gal})\dfrac{(3.9)}{100}\dfrac{(71)}{100}\dfrac{(56)}{100}}{35,000 \text{ cu ft}} = \dfrac{0.01 \text{ lbs/day VS Destroyed}}{\text{cu ft}}$

8. $\dfrac{153 \text{ mg/L}}{2260 \text{ mg/L}} = 0.07$

9. $(2130 \text{ mg/}L)(0.23 \text{ MG})(8.34 \text{ lbs/gal}) = 4086 \text{ lbs lime}$

10. $22 = \dfrac{x \text{ gal}}{(0.785)(45 \text{ ft})(45 \text{ ft})(20 \text{ ft})(7.48 \text{ gal/cu ft})} \times 100$

    $\dfrac{(22)(0.785)(45)(45)(20)(7.48 \text{ gal/cu ft})}{100} = x$

    $52,318 \text{ gal} = x$

11. $(4260 \text{ gpd Sludge})(8.34 \text{ lbs/gal})\dfrac{(5.3)}{100}\dfrac{(70)}{100} = 1318 \text{ lbs/day VS}$

12. $\dfrac{(2820 \text{ gpd})(8.34 \text{ lbs/gal})\dfrac{(5.7)}{100} + (4650 \text{ gpd})(8.34 \text{ lbs/gal})\dfrac{(3.6)}{100}}{(2820 \text{ gpd})(8.34 \text{ lbs/gal}) + (4650 \text{ gpd})(8.34 \text{ lbs/gal})} \times 100$

    $\dfrac{1341 \text{ lbs/day} + 1396 \text{ lbs/day}}{23,519 \text{ lbs/day} + 38,781 \text{ lbs/day}} \times 100$

    $= 4.4\%$

ACHIEVEMENT TEST 16—Cont'd

13. $\dfrac{160 \text{ mg}/L}{2490 \text{ mg}/L} = 0.06$

14. $\dfrac{(41{,}820 \text{ lbs/day Sludge})\frac{(5)}{100}\frac{(62)}{100}}{(90{,}000 \text{ gal})(8.34 \frac{\text{lbs}}{\text{gal}})\frac{(6.1)}{100}\frac{(56)}{100}} = \dfrac{0.05 \text{ lbs VS Added/day}}{\text{lb VS in Digester}}$

15. $(0.785)(0.83 \text{ ft})(0.83 \text{ ft})(0.33 \text{ ft})(7.48 \text{ gal/cu ft}) = 1.3 \text{ gal/stroke}$

   $(1.3 \text{ gal/stroke})(30 \text{ strokes/min}) = 39 \text{ gpm}$

16. $\dfrac{(19{,}790 \text{ gpd})(8.34 \frac{\text{lbs}}{\text{gal}})\frac{(6)}{100}\frac{(68)}{100}}{\dfrac{(0.785)(45 \text{ ft})(45 \text{ ft})(23 \text{ ft})}{1000}} = \dfrac{184 \text{ lbs VS Added/day}}{1000 \text{ cu ft}}$

17. $(2100 \text{ mg}/L)(0.2 \text{ MG})(8.34 \text{ lbs/gal}) = 3503 \text{ lbs Lime}$

18. $\dfrac{0.69 - 0.53}{0.69 - (0.69)(0.53)} \times 100 = 49\% \text{ Reduction}$

19. $\dfrac{6870 \text{ cu ft/day}}{560 \text{ lbs VS destroyed/day}} = \dfrac{12.3 \text{ cu ft Gas}}{\text{lb VS Destroyed}}$

20. $\dfrac{0.08 \text{ lb/day VS Added}}{1 \text{ lb VS in Dig.}} = \dfrac{(1170 \text{ gpd})(8.34 \text{ lbs/gal})\frac{(4.8)}{100}\frac{(68)}{100}}{(x \text{ gal})(8.5 \text{ lbs/gal})\frac{(7.2)}{100}\frac{(54)}{100}}$

   $x = \dfrac{(1170)(8.34)(0.048)(0.68)}{(0.08)(8.5)(0.072)(0.54)}$

   $x = 12{,}047 \text{ gal}$

21. $\dfrac{0.71 - 0.55}{0.71 - (0.71)(0.55)} \times 100 = 50\%$

22. $\dfrac{(0.785)(50 \text{ ft})(50 \text{ ft})(10 \text{ ft})(7.48 \text{ gal/cu ft})}{9500 \text{ gpd}} = 15.5 \text{ days}$

23. $\dfrac{21{,}300 \text{ cu ft gas/day}}{(2580 \text{ lbs VS/day})\frac{(58)}{100}} = 14 \text{ cu ft/lb VS Destroyed}$

24. $$\frac{\dfrac{(3100 \text{ gpd})(8.34 \text{ lbs/gal})\dfrac{(6.3)}{100}\dfrac{(66)}{100}\dfrac{(56)}{100}}{\dfrac{(0.785)(45 \text{ ft})(45 \text{ ft})(20 \text{ ft})}{1000}}} = \frac{18.9 \text{ lbs/day VS Destroyed}}{1000\text{-cu ft}}$$

25. <u>First calculate solids and water entering the digester:</u>

    • lbs/day total solids entering:

    (28,300 lbs/day)$\underset{\text{Sludge}}{}\dfrac{(5.9)}{100}$ = 1670 lbs/day Total Solids

    • lbs/day volatile solids entering:

    (1670 lbs/day)$\underset{\text{Solids}}{}\dfrac{(69)}{100}$ = 1152 lbs/day Volatile Solids

    • lbs/day fixed solids entering:

    $\underset{\text{Tot. Solids}}{1670 \text{ lbs/day}} - \underset{\text{Vol. Solids}}{1152 \text{ lbs/day}}$ = 518 lbs/day Fixed Solids

    • lbs/day water entering:

    $\underset{\text{Sludge}}{28,300 \text{ lbs/day}} - \underset{\text{Tot. Solids}}{1670 \text{ lbs/day}}$ = 26,630 lbs/day Water

<u>Then calculate solids, water and gas leaving the digester:</u>
(To begin these calculations, you must first determine the percent volatile solids reduction during digestion.)

    • *% Volatile Solids Reduction*

$$\underset{\text{Reduction}}{\% \text{ VS}} = \frac{\text{In} - \text{Out}}{\text{In} - (\text{In x Out})} \text{ x } 100$$

$$= \frac{0.69 - 0.52}{0.69 - (0.69)(0.52)} \text{ x } 100$$

$$= \frac{0.17}{0.3312} \text{ x } 100$$

$$= 51\% \text{ VS Reduction}$$

    • *lbs Gas Produced*
    (VS destroyed or reduced = lbs gas produced)

    (lbs VS Entering Digester) $\dfrac{(\% \text{ VS Reduction})}{100}$ = lbs Gas Produced

    (1152 lbs Vol. Sol.) $\dfrac{(51)}{100}$ = 588 lbs Gas Produced

(Answer Cont'd On Next Page)

ACHIEVEMENT TEST 16—Cont'd

- *lbs Volatile Solids in Digested Sludge*

    lbs VS Entering Digester − lbs VS Destroyed = lbs VS Leaving Digester

    $$1152 \text{ lbs} - 588 \text{ lbs} = 564 \text{ lbs VS in Digested Sludge}$$

- *lbs Total Solids in Digested Sludge*

    $$\frac{\text{lbs Vol. Sol.}}{\dfrac{\% \text{ VS}}{100}} = \text{lbs Total Solids}$$

    $$\frac{564 \text{ lbs VS}}{\dfrac{52}{100}} = 1085 \text{ lbs Total Solids}$$

- *lbs Fixed Solids in Digested Sludge*

    Total Solids, lbs − Volatile Solids, lbs = Fixed Solids, lbs

    $$1085 \text{ lbs} - 564 \text{ lbs} = 521 \text{ lbs}$$

- *lbs Digested Sludge*

    $$\frac{\text{lbs Total Solids}}{\dfrac{\% \text{ Solids}}{100}} = \text{lbs Digested Sludge}$$

    $$\frac{1085 \text{ lbs Tot. Sol.}}{\dfrac{4.0}{100}} = 27{,}125 \text{ lbs Digested Sludge}$$

- *lbs Water in Digested Sludge*

    Sludge, lbs − Total Solids, lbs = Water, lbs

    $$27{,}125 \text{ lbs} - 1085 \text{ lbs} = 26{,}040 \text{ lbs}$$

Summary Comparison	
Sludge Entering	**Sludge Leaving**

Tot. Sol.—1670 lbs
(VS—1152 lbs)
(FS— 518 lbs)
 	
Water—26,630 lbs
        28,300 lbs
	Tot. Sol.—1085 lbs
	(VS—564 lbs)
	(FS— 521 lbs)
	
	Water—26,040 lbs
	Gas— 588 lbs
	        27,713 lbs

26. $$\frac{0.04 \text{ cfm}}{1 \text{ cu ft Dig. Vol.}} = \frac{x \text{ cfm air required}}{(85 \text{ ft})(25 \text{ ft})(10 \text{ ft})}$$

$$(0.04)(85)(25)(10) = x$$

$$850 \text{ cfm} = x$$
$$\text{Air}$$

27. $(24 \text{ mg/L})(0.14 \text{ MG})(8.34 \text{ lbs/gal}) = 28 \text{ lbs Caustic}$

28. $$\frac{6.1 \text{ mg/L} - 3.9 \text{ mg/L}}{3 \text{ min}} \times \frac{60 \text{ min}}{\text{hr}} = 44.0 \text{ mg/L/hr}$$

29. First calculate min/day operation:

$$\frac{(122 \text{ mg/L})(2.4 \text{ MGD})(8.34 \text{ lbs})}{\underset{100}{3.1}} = (25 \text{ gpm})(8.34 \frac{\text{lbs}}{\text{gal}})(x \frac{\text{min}}{\text{day}})$$

$$\frac{(122)(2.4)(8.34)}{(0.031)(25)(8.34)} = x$$

$$378 \text{ min/day} = x$$

Then convert min/day to min/hr:

$$\frac{378 \text{ min/day}}{24 \text{ hrs/day}} = 16 \text{ min/hr}$$
$$\text{Pump Operating Time}$$

30. First calculate detention time for the sludge containing 2.4% solids:

$$\frac{(0.785)(30 \text{ ft})(30 \text{ ft})(22 \text{ ft})(7.48 \text{ gal/cu ft})}{10,000 \text{ gpd}} = 11.6 \text{ days}$$

Then determine the detention time for the sludge containing 4.5% solids:

$$\frac{(0.785)(30 \text{ ft})(30 \text{ ft})(22 \text{ ft})(7.48 \text{ gal/cu ft})}{5300 \text{ gpd}} = 21.9 \text{ days}$$

Chapter 17

PRACTICE PROBLEMS 17.1

1. $$\frac{\frac{(1000 \text{ gal})}{3 \text{ hr}} (8.34 \frac{\text{lbs}}{\text{gal}}) \frac{(3.6)}{100}}{130 \text{ sq ft}} = 0.8 \text{ lbs/hr/sq ft}$$

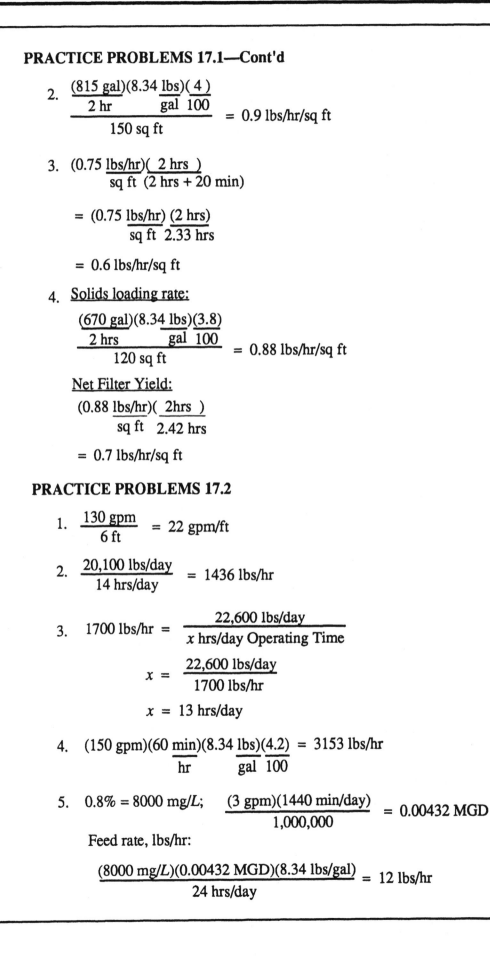

PRACTICE PROBLEMS 17.1—Cont'd

2. $\dfrac{\dfrac{(815 \text{ gal})(8.34 \dfrac{\text{lbs}}{\text{gal}})(\dfrac{4}{100})}{2 \text{ hr}}}{150 \text{ sq ft}} = 0.9 \text{ lbs/hr/sq ft}$

3. $(0.75 \dfrac{\text{lbs/hr}}{\text{sq ft}})(\dfrac{2 \text{ hrs}}{2 \text{ hrs} + 20 \text{ min}})$

   $= (0.75 \dfrac{\text{lbs/hr}}{\text{sq ft}}) (\dfrac{2 \text{ hrs}}{2.33 \text{ hrs}})$

   $= 0.6 \text{ lbs/hr/sq ft}$

4. Solids loading rate:

   $\dfrac{\dfrac{(670 \text{ gal})(8.34 \dfrac{\text{lbs}}{\text{gal}})(\dfrac{3.8}{100})}{2 \text{ hrs}}}{120 \text{ sq ft}} = 0.88 \text{ lbs/hr/sq ft}$

   Net Filter Yield:

   $(0.88 \dfrac{\text{lbs/hr}}{\text{sq ft}})(\dfrac{2 \text{hrs}}{2.42 \text{ hrs}})$

   $= 0.7 \text{ lbs/hr/sq ft}$

PRACTICE PROBLEMS 17.2

1. $\dfrac{130 \text{ gpm}}{6 \text{ ft}} = 22 \text{ gpm/ft}$

2. $\dfrac{20,100 \text{ lbs/day}}{14 \text{ hrs/day}} = 1436 \text{ lbs/hr}$

3. $1700 \text{ lbs/hr} = \dfrac{22,600 \text{ lbs/day}}{x \text{ hrs/day Operating Time}}$

   $x = \dfrac{22,600 \text{ lbs/day}}{1700 \text{ lbs/hr}}$

   $x = 13 \text{ hrs/day}$

4. $(150 \text{ gpm})(\dfrac{60 \text{ min}}{\text{hr}})(8.34 \dfrac{\text{lbs}}{\text{gal}})(\dfrac{4.2}{100}) = 3153 \text{ lbs/hr}$

5. $0.8\% = 8000 \text{ mg/}L$; $\dfrac{(3 \text{ gpm})(1440 \text{ min/day})}{1,000,000} = 0.00432 \text{ MGD}$

   Feed rate, lbs/hr:

   $\dfrac{(8000 \text{ mg/}L)(0.00432 \text{ MGD})(8.34 \text{ lbs/gal})}{24 \text{ hrs/day}} = 12 \text{ lbs/hr}$

6. First calculate tons/hr solids loading rate:

$$\frac{3153 \text{ lbs solids/hr}}{2000 \text{ lbs/ton}} = 1.6 \text{ tons solids/hr}$$

$$\frac{\text{Flocculant}}{\text{Dosage, lbs/ton}} = \frac{12 \text{ lbs flocculant/hr}}{1.6 \text{ tons solids/hr}}$$

$$= 7.5 \text{ lbs/tons}$$

7. $\text{\% Recovery} = \dfrac{(15)(3.7 - 0.039)}{(3.7)(15 - 0.039)} \times 100$

$$= \frac{54.9}{55.4} \times 100$$

$$= 99\% \text{ Recovery}$$

PRACTICE PROBLEMS 17.3

1. $\dfrac{(85 \text{ gpm})(60 \frac{\text{min}}{\text{hr}})(8.34 \frac{\text{lbs}}{\text{gal}})(\frac{4.8}{100})}{310 \text{ sq ft}} = 6.6 \text{ lbs/hr/sq ft}$

2. $\dfrac{(6750 \text{ lbs/hr})(\frac{28}{100})}{300 \text{ sq ft}} = 6.3 \text{ lbs/hr/sq ft}$

3. $3.2 \text{ lbs/hr/sq ft} = \dfrac{\dfrac{5300 \text{ lbs/day}}{x \text{ hrs/day Oper.}}}{220 \text{ sq ft}} \dfrac{(95)}{100}$

$$3.2 \text{ lbs/hr/sq ft} = \frac{5300 \text{ lbs/day}}{x \text{ hrs/day}} \frac{(1)}{220 \text{ sq ft}} \frac{(95)}{100}$$

$$x = \frac{(5300)(1)(95)}{(3.2)(220)(100)}$$

$$= \boxed{\begin{array}{c} 7.2 \text{ hrs/day} \\ \text{Operation} \end{array}}$$

PRACTICE PROBLEMS 17.3—Cont'd

4.
$$\text{Filter Yield, lbs/hr/sq ft} = \frac{\dfrac{17{,}240 \text{ lbs/day}}{10 \text{ hrs/day}}}{280 \text{ sq ft}} \; \frac{(93)}{100}$$

$$= \frac{(1724 \text{ lbs/hr})}{280 \text{ sq ft}} \; \frac{(93)}{100}$$

$$= \; 5.7 \text{ lbs/hr/sq ft}$$

5.
$$\text{\% Solids Recovery} = \frac{\dfrac{(18{,}300 \text{ lbs/hr})(25)}{100}}{\dfrac{(84{,}800 \text{ lbs/hr})(5.7)}{100}} \times 100$$

$$= \frac{4575 \text{ lbs/hr}}{4834 \text{ lbs/hr}} \times 100$$

$$= \; 95\% \text{ Solids Recovery}$$

PRACTICE PROBLEMS 17.4

1. (200 ft) (20 ft) (0.5 ft) (7.48 gal/cu ft) = 14,960 gal

2. $(230 \text{ ft}) (25 \text{ ft}) \dfrac{(8 \text{ in.})}{12 \text{ in./ft}} (7.48 \text{ gal/cu ft}) = 28{,}673 \text{ gal}$

3. $\dfrac{\dfrac{(166{,}000 \text{ lbs})}{22 \text{ days}} \dfrac{(365 \text{ days})}{\text{yr}} \dfrac{(4.8)}{100}}{(180 \text{ ft}) (25 \text{ ft})} = 29.4 \text{ lbs/yr/sq ft}$

4. First calculate the lbs of sludge applied:

$(210 \text{ ft})(25 \text{ ft})\dfrac{(10 \text{ in.})}{12 \text{ in./ft}}\dfrac{(7.48 \text{ gal})}{\text{cu ft}}\dfrac{(8.34 \text{ lbs})}{\text{gal}} = 272{,}927 \text{ lbs}$

Then determine the solids loading rate:

$$\text{Solids Loading Rate, lbs/yr/sq ft} = \frac{\dfrac{(272{,}927 \text{ lbs})}{25 \text{ days}} \dfrac{(365 \text{ days})}{\text{yr}} \dfrac{(3.8)}{100}}{(210 \text{ ft})(25 \text{ ft})}$$

$$= 28.8 \text{ lbs/yr/sq ft}$$

5. (0.785)(40 ft)(40 ft)(2.5 ft) = 3140 cu ft

6. a) $(0.785)(45 \text{ ft})(45 \text{ ft})(1.33 \text{ ft}) = (80 \text{ ft})(40 \text{ ft})(x \text{ ft})$

   $x = 0.66 \text{ ft}$

   b) $(0.66 \text{ ft})(12 \text{ in./ft}) = 8 \text{ inches}$

PRACTICE PROBLEMS 17.5

1. Sludge moisture is 78%

$$\frac{(4500 \text{ lbs/day})\frac{(78)}{100} + (3700 \text{ lbs/day})\frac{(25)}{100}}{4500 \text{ lbs/day} + 3700 \text{ lbs/day}} \times 100$$

$$= \frac{3510 \text{ lbs/day} + 925 \text{ lbs/day}}{8200 \text{ lbs/day}} \times 100$$

$$= \boxed{54\% \text{ Moisture}}$$

2. Sludge moisture is 85%

$$\frac{(9,600 \text{ lbs/day})\frac{(85)}{100} + (10,500 \text{ lbs/day})\frac{(28)}{100}}{14,600 \text{ lbs/day} + 10,500 \text{ lbs/day}} \times 100$$

$$= \frac{8160 \text{ lbs/day} + 2940 \text{ lbs/day}}{25,100 \text{ lbs/day}} \times 100$$

$$= \boxed{44\% \text{ Moisture}}$$

3. Sludge moisture = 82%

$$45 = \frac{(4,600 \text{ lbs/day})\frac{(82)}{100} + (x \text{ lbs/day})\frac{(28)}{100}}{4,600 \text{ lbs/day} + x \text{ lbs/day}} \times 100$$

$$\frac{45}{100} = \frac{3772 \text{ lbs/day} + (x \text{ lbs/day})(0.28)}{4600 \text{ lbs/day} + x \text{ lbs/day}}$$

$$0.45 (4600 + x) = 3772 + 0.28x$$

$$2070 + 0.45 x = 3772 + 0.28x$$

$$0.45 x - 0.28x = 3772 - 2070$$

$$0.17x = 1702$$

$$x = \boxed{10,012 \text{ lbs/day}}$$

PRACTICE PROBLEMS 17.5—Cont'd

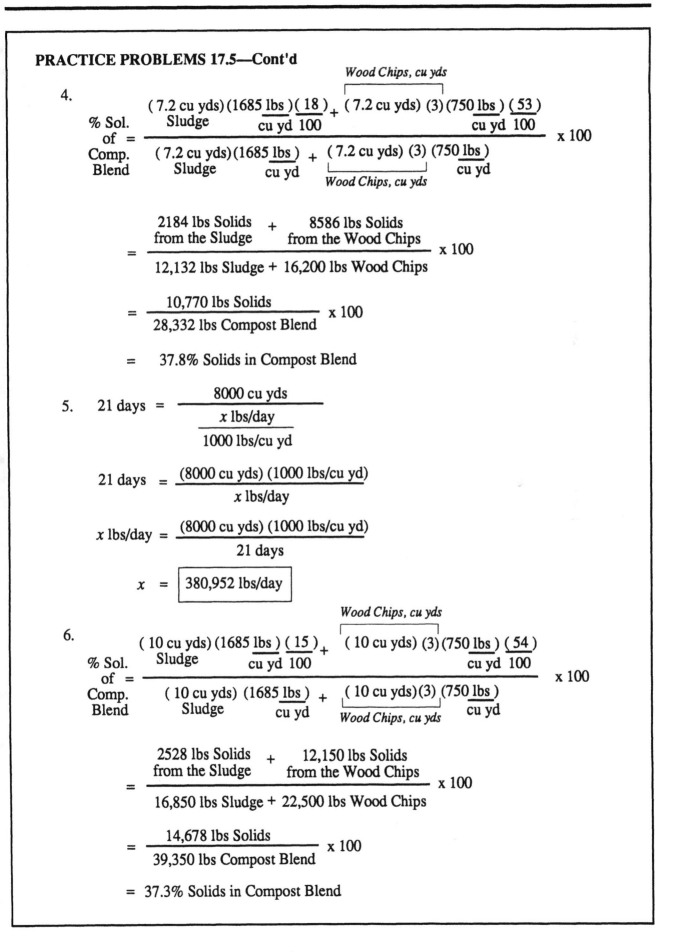

4.

$$\text{% Sol. of Comp. Blend} = \frac{(7.2 \text{ cu yds})(1685 \frac{\text{lbs}}{\text{cu yd}})(\frac{18}{100}) + \overbrace{(7.2 \text{ cu yds})(3)(750 \frac{\text{lbs}}{\text{cu yd}})(\frac{53}{100})}^{\text{Wood Chips, cu yds}}}{(7.2 \text{ cu yds})(1685 \frac{\text{lbs}}{\text{cu yd}}) + \underbrace{(7.2 \text{ cu yds})(3)(750 \frac{\text{lbs}}{\text{cu yd}})}_{\text{Wood Chips, cu yds}}} \times 100$$

$$= \frac{\underset{\text{from the Sludge}}{2184 \text{ lbs Solids}} + \underset{\text{from the Wood Chips}}{8586 \text{ lbs Solids}}}{12{,}132 \text{ lbs Sludge} + 16{,}200 \text{ lbs Wood Chips}} \times 100$$

$$= \frac{10{,}770 \text{ lbs Solids}}{28{,}332 \text{ lbs Compost Blend}} \times 100$$

$$= 37.8\% \text{ Solids in Compost Blend}$$

5. $$21 \text{ days} = \frac{8000 \text{ cu yds}}{\dfrac{x \text{ lbs/day}}{1000 \text{ lbs/cu yd}}}$$

$$21 \text{ days} = \frac{(8000 \text{ cu yds})(1000 \text{ lbs/cu yd})}{x \text{ lbs/day}}$$

$$x \text{ lbs/day} = \frac{(8000 \text{ cu yds})(1000 \text{ lbs/cu yd})}{21 \text{ days}}$$

$$x = \boxed{380{,}952 \text{ lbs/day}}$$

6.

$$\text{% Sol. of Comp. Blend} = \frac{(10 \text{ cu yds})(1685 \frac{\text{lbs}}{\text{cu yd}})(\frac{15}{100}) + \overbrace{(10 \text{ cu yds})(3)(750 \frac{\text{lbs}}{\text{cu yd}})(\frac{54}{100})}^{\text{Wood Chips, cu yds}}}{(10 \text{ cu yds})(1685 \frac{\text{lbs}}{\text{cu yd}}) + \underbrace{(10 \text{ cu yds})(3)(750 \frac{\text{lbs}}{\text{cu yd}})}_{\text{Wood Chips, cu yds}}} \times 100$$

$$= \frac{\underset{\text{from the Sludge}}{2528 \text{ lbs Solids}} + \underset{\text{from the Wood Chips}}{12{,}150 \text{ lbs Solids}}}{16{,}850 \text{ lbs Sludge} + 22{,}500 \text{ lbs Wood Chips}} \times 100$$

$$= \frac{14{,}678 \text{ lbs Solids}}{39{,}350 \text{ lbs Compost Blend}} \times 100$$

$$= 37.3\% \text{ Solids in Compost Blend}$$

7. $\dfrac{21}{\text{days}} = \dfrac{(7750 \text{ cu yds})(1000 \text{ lbs/cu yd})}{\dfrac{x \text{ lbs/day}}{\substack{\text{Dry Solids} \\ 0.18}} + \dfrac{(x \text{ lbs/day})}{\substack{\text{Dry Solids} \\ 0.18}}\dfrac{(3 \text{ Mix Ratio})}{1}\dfrac{(750 \text{ lbs/cu yd})}{1685 \text{ lbs/cu yd}}}$

First simplify terms, as possible:

$$21 = \dfrac{7{,}750{,}000}{\dfrac{x}{0.18} + 7.42\,x}$$

The x term in the denominator can be further simplified:

$$28 = \dfrac{7{,}750{,}000}{\dfrac{1}{0.18}\,x + 7.42\,x}$$

$$21 = \dfrac{7{,}750{,}000}{5.56\,x + 7.42\,x}$$

$$21 = \dfrac{7{,}750{,}000}{12.98\,x}$$

$$12.98\,x = \dfrac{7{,}750{,}000}{21}$$

$$x = \dfrac{7{,}750{,}000}{(21)(12.98)}$$

$$x = 28{,}432 \text{ lbs/day Dry Solids}$$

8.   a) 35.6% Dry Solids

b) 60 dry tons/wk

9.   a) 2.8 mix ratio

b) 75 dry tons/wk

CHAPTER 17—ACHIEVEMENT TEST

1. $(140 \text{ gpm})(60 \frac{\text{min}}{\text{hr}}) (8.34 \frac{}{\text{lbs/gal}}) (\frac{4.6}{100}) = 3223 \text{ lbs/hr}$

2. $\dfrac{(24,100 \text{ lbs/day})}{14 \text{ hrs/day}} = 1721 \text{ lbs/hr}$

3. $\dfrac{\dfrac{(750 \text{ gal})}{2 \text{ hrs}} (8.34 \frac{\text{lbs}}{\text{gal}}) (\frac{3.8}{100})}{135 \text{ sq ft}} = 0.9 \text{ lbs/hr/sq ft}$

4. First, calculate the solids loading rate in lbs/hr:

   $(160 \text{ gpm})(60 \frac{\text{min}}{\text{hr}}) (8.34 \frac{\text{lbs}}{\text{gal}}) (\frac{4}{100}) = 3203 \text{ lbs/hr}$

   Expressed as tons/hr, this is:

   $\dfrac{3203 \text{ lbs/hr}}{2000 \text{ lbs/ton}} = 1.6 \text{ tons/hr}$

   Next, calculate the flocculant feed in lbs/day then lbs/hr: (2.5 gpm must be converted to MGD flow rate to use the mg/*L* to lbs/day equation):

   $\dfrac{(2.5 \text{ gpm})(1440 \text{ min/day})}{1,000,000} = 0.0036 \text{ MGD}$

   $(8000 \text{ mg/}L)(0.0036 \text{ MGD})(8.34 \text{ lbs/gal}) = 240 \text{ lbs/day}$

   Expressed as lbs/hr, this feed rate is:

   $\dfrac{240 \text{ lbs/day}}{24 \text{ hrs/day}} = 10 \text{ lbs/hr Flocculant}$

   Now the flocculant dose, in lbs flocculant/ton solids, can be determined:

   $\dfrac{\text{Flocculant Dose}}{\text{lbs/ton}} = \dfrac{10 \text{ lbs/hr Flocculant}}{1.6 \text{ tons/hr Solids}}$

   $= 6.3 \text{ lbs Flocculant/ton Solids}$

5. $\text{Yield} = (0.7 \frac{\text{lbs/hr}}{\text{sq ft}}) (\frac{2 \text{ hrs}}{2 \text{ hrs} + 22 \text{ min}})$

   $= (0.7 \frac{\text{lbs/hr}}{\text{sq ft}}) (\frac{2 \text{ hrs}}{2.37 \text{ hrs}})$

   $= 0.6 \text{ lbs/hr/sq ft}$

6. $23{,}100 \text{ mg/}L - 720 \text{ mg/}L = 22{,}380 \text{ mg/}L$

7. $\dfrac{(75 \text{ gpm})(60 \frac{\text{min}}{\text{hr}})(8.34 \frac{\text{lbs}}{\text{gal}})(\frac{5.3}{100})}{300 \text{ sq ft}} = 6.6 \text{ lbs/hr/sq ft}$

8. $\dfrac{(7400 \text{ lbs/hr})(\frac{25}{100})}{310 \text{ sq ft}} = 6.0 \text{ lbs/hr/sq ft}$

9. $\dfrac{(15)(2.6 - 0.047)}{(2.6)(15 - 0.047)} \times 100 = \dfrac{38.295}{38.8778} \times 100$

    $= 99\% \text{ Recovery}$

10. $1900 \text{ lbs/hr} = \dfrac{26{,}700 \text{ lbs/day}}{x \text{ hrs/day}}$

    $x = \dfrac{26{,}700 \text{ lbs/day}}{1900 \text{ lbs/hr}}$

    $= 14 \text{ hrs/day}$

11. $2.9 = \dfrac{\dfrac{5600 \text{ lbs/day}}{x \text{ hrs/day}}}{250 \text{ sq ft}} \dfrac{(94)}{100}$

    $2.9 = \dfrac{(5600 \text{ lbs/day})}{x \text{ lbs/day}} \dfrac{(1)}{250 \text{ sq ft}} \dfrac{(94)}{100}$

    $x = \dfrac{(5600)(1)(94)}{(2.9)(250)(100)}$

    $x = 7.3 \text{ hrs/day Operation}$

12. $(210 \text{ ft})(25 \text{ ft})(\dfrac{8 \text{ in.}}{12 \text{ in./ft}})(7.48 \text{ gal/cu ft}) = 26{,}180 \text{ gallons}$

13. $\dfrac{(14{,}100 \frac{\text{lbs/hr}}{\text{wet cake}})(\frac{30}{100})}{(89{,}000 \frac{\text{lbs/hr}}{\text{sludge}})(\frac{5.1}{100})} \times 100$

    $= \dfrac{4230 \text{ lbs/hr}}{4539 \text{ lbs/hr}} \times 100 = 93\% \text{ Solids Recovery}$

CHAPTER 17—ACHIEVEMENT TEST—Cont'd

14. First calculate the lbs of sludge applied: (6 in. = 0.5 ft)

$$(190 \text{ ft})(20 \text{ ft})(0.5 \text{ ft})(7.48 \frac{\text{gal}}{\text{cu ft}})(8.34 \frac{\text{lbs}}{\text{gal}}) = 118,528 \text{ lbs Sludge}$$

Then determine the solids loading rate:

$$\frac{\dfrac{(118,528 \text{ lbs})(365 \frac{\text{days}}{\text{yr}})(\frac{4.8}{100})}{20 \text{ days}}}{(190 \text{ ft})(20 \text{ ft})} = \frac{103,831 \text{ lbs/yr}}{3800 \text{ sq ft}}$$

$$= 27 \text{ lbs/yr/sq ft}$$

15. (a) $(0.785)(45 \text{ ft})(45 \text{ ft})(1 \text{ ft}) = (180 \text{ ft})(25 \text{ ft})(x \text{ ft})$

$$\frac{(0.785)(45)(45)(1)}{(180)(25)} = x$$

$$0.35 \text{ ft} = x$$

(b) Convert 0.35 ft to inches:

$$(0.35 \text{ ft})(12 \frac{\text{in.}}{\text{ft}}) = 4 \text{ inches}$$

16. The dewatered sludge has a solids content of 25%; therefore the moisture content is 75%.

$$50 = \frac{(6600 \text{ lbs/day})(\frac{75}{100}) + (x \text{ lbs/day})(\frac{35}{100})}{6600 \text{ lbs/day} + x \text{ lbs/day}} \times 100$$

$$\frac{50}{100} = \frac{4950 \text{ lbs/day} + (x \text{ lbs/day})(0.35)}{6600 \text{ lbs/day} + x \text{ lbs/day}}$$

$$0.5 (6600 + x) = 4950 + 0.35x$$

$$3300 + 0.5 x = 4950 + 0.35x$$

$$0.5x - 0.35 x = 4950 - 3300$$

$$0.15 x = 1650$$

$$x = 11,000 \text{ lbs/day}$$
$$\text{Compost Req'd}$$

17.

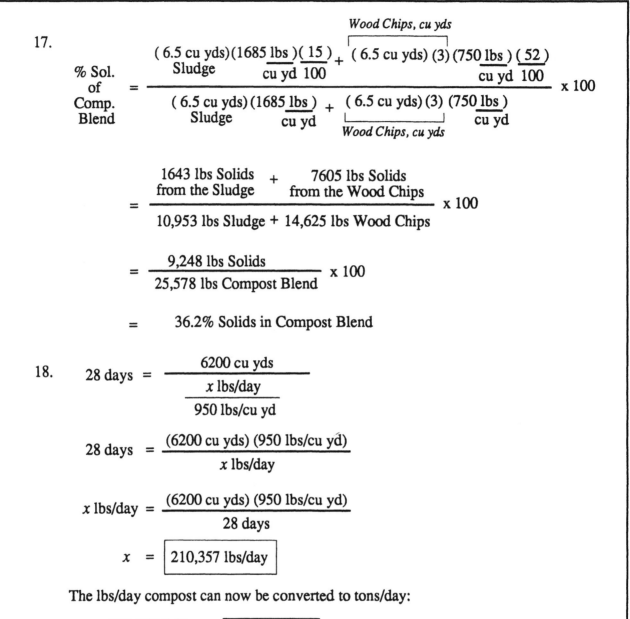

$$\text{\% Sol. of Comp. Blend} = \frac{(6.5 \text{ cu yds})(1685 \tfrac{\text{lbs}}{\text{cu yd}})(\tfrac{15}{100})_{\text{Sludge}} + (6.5 \text{ cu yds})(3)(750 \tfrac{\text{lbs}}{\text{cu yd}})(\tfrac{52}{100})}{(6.5 \text{ cu yds})(1685 \tfrac{\text{lbs}}{\text{cu yd}})_{\text{Sludge}} + (6.5 \text{ cu yds})(3)(750 \tfrac{\text{lbs}}{\text{cu yd}})} \times 100$$

$$= \frac{1643 \text{ lbs Solids from the Sludge} + 7605 \text{ lbs Solids from the Wood Chips}}{10{,}953 \text{ lbs Sludge} + 14{,}625 \text{ lbs Wood Chips}} \times 100$$

$$= \frac{9{,}248 \text{ lbs Solids}}{25{,}578 \text{ lbs Compost Blend}} \times 100$$

$$= 36.2\% \text{ Solids in Compost Blend}$$

18.

$$28 \text{ days} = \frac{6200 \text{ cu yds}}{\dfrac{x \text{ lbs/day}}{950 \text{ lbs/cu yd}}}$$

$$28 \text{ days} = \frac{(6200 \text{ cu yds})(950 \text{ lbs/cu yd})}{x \text{ lbs/day}}$$

$$x \text{ lbs/day} = \frac{(6200 \text{ cu yds})(950 \text{ lbs/cu yd})}{28 \text{ days}}$$

$$x = \boxed{210{,}357 \text{ lbs/day}}$$

The lbs/day compost can now be converted to tons/day:

$$\frac{210{,}357 \text{ lb/day}}{2{,}000 \text{ lbs/ton}} = \boxed{105 \text{ tons/day}}$$

CHAPTER 17—ACHIEVEMENT TEST—Cont'd

19. $\dfrac{25 \text{ days}}{} = \dfrac{(8000 \text{ cu yds}) (1000 \text{ lbs/cu yd})}{\dfrac{x \text{ lbs/day Dry Solids}}{0.18} + \dfrac{x \text{ lbs/day Dry Solids}}{0.18} \left(\dfrac{3.3 \text{ Mix Ratio}}{1}\right) \dfrac{(750 \text{ lbs/cu yd})}{1685 \text{ lbs/cu yd}}}$

First simplify terms, as possible:

$$25 = \dfrac{8{,}000{,}000}{\dfrac{x}{0.18} + 8.16\,x}$$

The x term in the denominator can be further simplified:

$$25 = \dfrac{8{,}000{,}000}{\dfrac{1}{0.18}\,x + 8.16\,x}$$

$$25 = \dfrac{8{,}000{,}000}{5.56\,x + 8.16\,x}$$

$$25 = \dfrac{8{,}000{,}000}{13.72\,x}$$

$$13.72\,x = \dfrac{8{,}000{,}000}{25}$$

$$x = \dfrac{8{,}000{,}000}{(25)\,(13.72)}$$

$$x = \boxed{23{,}324 \text{ lbs/day Dry Solids}}$$

20. 39% total solids compost
    87 dry tons/wk

Chapter 18

PRACTICE PROBLEMS 18.1

1. $\dfrac{6 \text{ mg/L} - 3.5 \text{ mg/L}}{\dfrac{5 \text{ m}L}{300 \text{ m}L}} = \boxed{150 \text{ mg/L BOD}}$

2. $\dfrac{8.3 \text{ mg/L} - 4.2 \text{ mg/L}}{\dfrac{10 \text{ m}L}{300 \text{ m}L}} = \boxed{123 \text{ mg/L BOD}}$

PRACTICE PROBLEMS 18.1—Cont'd

3. $\dfrac{190 + 198 + 205 + 202 + 210 + 201 + 197}{7} = \dfrac{1403}{7} = \boxed{200 \text{ mg/L}}$

4. 7-day average for August 9th: (Aug. 9th + 6 previous days)

$$\begin{array}{l}\text{7-day Aver.} \\ \text{BOD}\end{array} = \dfrac{208 + 203 + 215 + 220 + 225 + 206 + 195}{7} = \dfrac{1472}{7}$$

$$= \boxed{210 \text{ mg/L}}$$

7-day average for August 10th: (Use the shortcut method for determining the new total)

$$\begin{array}{l}\text{7-day Aver.} \\ \text{BOD}\end{array} = \dfrac{1472 - 208 + 198}{7} = \dfrac{1462}{7}$$

$$= \boxed{209 \text{ mg/L}}$$

7-day average for August 11th: (Use the shortcut method for determining the new total)

$$\begin{array}{l}\text{7-day Aver.} \\ \text{BOD}\end{array} = \dfrac{1462 - 203 + 210}{7} = \dfrac{1469}{7}$$

$$= \boxed{210 \text{ mg/L}}$$

PRACTICE PROBLEMS 18.2

1. $\dfrac{2.7 \text{ moles}}{0.6 \text{ liters}} = \boxed{4.5 \text{ M}}$

2. $1.5 \text{ M} = \dfrac{x \text{ moles}}{0.8 \text{ liters}}$

   $x = 1.2 \text{ moles}$

3. $\dfrac{26 \text{ grams}}{40 \text{ grams/mole}} = 0.7 \text{ moles}$

4. $\dfrac{0.3 \text{ moles}}{1.6 \text{ liters}} = 0.2 \text{ M}$

5. $\begin{array}{r} 22.997 \\ 16.000 \\ \underline{1.008} \\ 40.005 \text{ grams} \end{array}$

PRACTICE PROBLEMS 18.3

1. $\dfrac{2.1 \text{ Equivalents}}{1.6 \text{ liters}} = \boxed{1.3 \text{ N}}$

2. $\dfrac{1.4 \text{ Equivalents}}{0.3 \text{ liters}} = \boxed{4.7 \text{ N}}$

3. $(0.7)(x \text{ mL}) = (0.05)(750 \text{ mL})$

$$x = \frac{(0.05)(750)}{0.7}$$

$$x = 54 \text{ ml NaOH}$$

4. $\dfrac{40 \text{ atomic wt.}}{2 \text{ valence electrons}} = 20$

5. $\dfrac{105 \text{ grams}}{53 \text{ grams/equivalent}} = 2.0 \text{ Equivalents}$

PRACTICE PROBLEMS 18.4

1. $\dfrac{410 \text{ mL}}{2000 \text{ mL}} \times 100 = 21\%$

2. $\dfrac{315 \text{ mL}}{2000 \text{ mL}} \times 100 = 16\%$

3. $\dfrac{390 \text{ mL}}{2000 \text{ mL}} \times 100 = 20\%$

4. $\dfrac{360 \text{ mL}}{2000 \text{ mL}} \times 100 = 18\%$

PRACTICE PROBLEMS 18.5

1. $\dfrac{15.9 \text{ mL/L Removed}}{16.5 \text{ mL/L in Influent}} \times 100 = 96\% \text{ Removed}$

2. $\dfrac{17.1 \text{ mL/L Removed}}{18 \text{ mL/L in Influent}} \times 100 = 95\% \text{ Removed}$

3. $\dfrac{19.2 \text{ m}L/L \text{ Removed}}{20 \text{ m}L/L \text{ in Influent}} \times 100 = 96\%$

4. $95 = \dfrac{x \text{ m}L/L \text{ Removed}}{19 \text{ m}L/L \text{ in Influent}} \times 100$

$x = \dfrac{(95)(19)}{100}$

$x = 18 \text{ m}L/L$

PRACTICE PROBLEMS 18.6

1.
Sludge Sample	Total Solids	Volatile Solids
84.15 g	25.17 g	25.17 g
− 22.40 g	− 22.40 g	− 23.29 g
61.75 g	2.77 g	1.88 g

a) $\begin{array}{l}\% \text{ Total} \\ \text{Solids}\end{array} = \dfrac{2.77 \text{ grams Tot. Sol.}}{61.75 \text{ grams Sludge}} \times 100$

$= 4.5\% \text{ Total Solids}$

b) $\begin{array}{l}\% \text{ Volatile} \\ \text{Solids}\end{array} = \dfrac{1.88 \text{ grams Vol. Sol.}}{2.77 \text{ grams Tot. Sol.}} \times 100$

$= 68\% \text{ Volatile Solids}$

2.
Sludge Sample	Total Solids	Volatile Solids
75.86 g	22.97 g	22.97 g
− 21.07 g	− 21.07 g	− 21.67 g
54.79 g	1.90 g	1.30 g

a) $\begin{array}{l}\% \text{ Total} \\ \text{Solids}\end{array} = \dfrac{1.90 \text{ grams Tol. Sol.}}{54.79 \text{ grams Sludge}} \times 100$

$= 3.5\% \text{ Total Solids}$

b) $\begin{array}{l}\% \text{ Volatile} \\ \text{Solids}\end{array} = \dfrac{1.30 \text{ grams Vol. Sol.}}{1.90 \text{ grams Tot. Sol.}} \times 100$

$= 68\% \text{ Volatile Solids}$

PRACTICE PROBLEMS 18.6—Cont'd

3. (a)

Total Solids	Volatile Solids
22.0173	22.0173
− 22.0024	− 22.0070
0.0149	0.0103

$$\frac{\%\ \text{Volatile}}{\text{Solids}} = \frac{0.0103\ \text{Vol. Sol.}}{0.0149\ \text{Tot. Sol.}} \times 100$$

$$= 69\%$$

(b) $\dfrac{0.0103\ \text{g VS}}{100\ \text{m}L} \times \dfrac{1000\ \text{mg}}{1\ \text{g}} = \dfrac{10.3\ \text{mg VS}}{100\ \text{m}L} \times \dfrac{\times 10}{\times 10} = 103\ \text{mg}/L\ \text{VS}$

PRACTICE PROBLEMS 18.7

1. (a) grams SS

25.6715 g SS &Dish
− 25.6670 g Dish
0.0045 g SS

$\dfrac{(0.0045\ \text{g})}{50\ \text{m}L} \dfrac{(1000\ \text{mg})}{1\ \text{g}} \dfrac{(20)}{20} = 90\ \text{mg}/L\ \text{SS}$

(b) grams VSS

25.6715 g Before Burning
− 25.6701 g After Burning
0.0014 g VSS

$\%\ \text{VSS} = \dfrac{0.0014}{0.0045} \times 100$

$= 31\%\ \text{VSS}$

2. (a) grams SS

36.1544 g SS &Dish
− 36.1477 g Dish
0.0067 g SS

$\dfrac{(0.0067\ \text{g})}{25\ \text{m}L} \dfrac{(1000\ \text{mg})}{1\ \text{g}} \dfrac{(40)}{40} = 268\ \text{mg}/L\ \text{SS}$

(b) grams VSS

36.1544 g Before Burning
− 36.1500 g After Burning
0.0044 g VSS

$\%\ \text{VSS} = \dfrac{0.0044\ \text{g}}{0.0067\ \text{g}} \times 100$

$= 66\%\ \text{VSS}$

3. (a) grams SS

28.3169 g SS &Dish
− 28.2986 g Dish
0.0183 g SS

$\dfrac{(0.0183\ \text{g})}{25\ \text{m}L} \dfrac{(1000\ \text{mg})}{1\ \text{g}} \dfrac{(40)}{40} = 732\ \text{mg}/L\ \text{SS}$

(b) grams VSS

28.3169 g Before Burning
− 28.3034 g After Burning
0.0135 g VSS

$\%\ \text{VSS} = \dfrac{0.0135\ \text{g}}{0.0183\ \text{g}} \times 100$

$= 74\%\ \text{VSS}$

PRACTICE PROBLEMS 18.8

1. $\dfrac{215\text{ m}L}{2180\text{ mg}}$ or $\dfrac{215\text{ m}L}{2.18\text{ g}}$ = 99 SVI

2. If 380 mL settled in 2 liters, then use 190 mL for 1 liter:

   $\dfrac{190\text{ m}L}{2260\text{ mg}}$ or $\dfrac{190\text{ m}L}{2.26\text{ g}}$ = 84 SVI

3. $\dfrac{2050\text{ m}L}{219\text{ mg}}$ x 100 = $\dfrac{2.05\text{ m}L}{219\text{ g}}$ x 100 = 0.94 SDI

4. $\dfrac{2050\text{ m}L}{182\text{ mg}}$ x 100 = $\dfrac{2.05\text{ m}L}{182\text{ g}}$ x 100 = 1.13 SDI

5. $\dfrac{205\text{ m}L}{2470\text{ mg}}$ or $\dfrac{205\text{ m}L}{2.47\text{ g}}$ = 83 SVI

PRACTICE PROBLEMS 18.9

1. 72° F + 40° = 112°
   And $\dfrac{(5)(112°)}{9}$ = 62°
   Then 62° − 40° = 22° C

2. 56° F + 40° = 96°
   And $\dfrac{(5)(96°)}{9}$ = 53°
   Then 53° − 40° = 13° C

3. 22° C + 40° = 62°
   And $\dfrac{(9)(62°)}{5}$ = 112°
   Then 112° − 40° = 72° F

4. 15° C + 40° = 55°
   And $\dfrac{(9)(55°)}{5}$ = 99°
   Then 99° − 40° = 59° F

CHAPTER 18—ACHIEVEMENT TEST

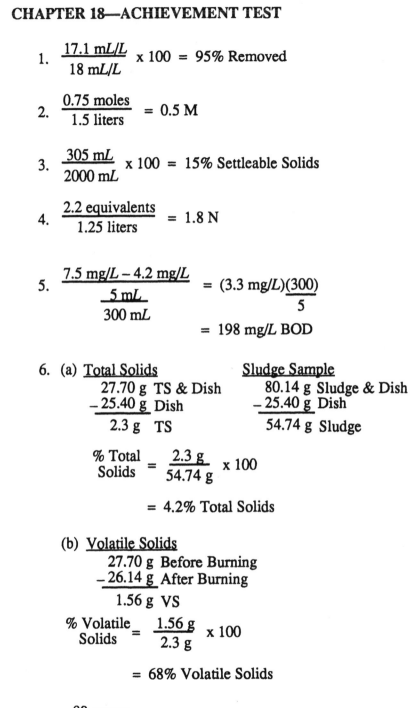

1. $\dfrac{17.1\ \text{mL/L}}{18\ \text{mL/L}} \times 100 = 95\%$ Removed

2. $\dfrac{0.75\ \text{moles}}{1.5\ \text{liters}} = 0.5\ \text{M}$

3. $\dfrac{305\ \text{mL}}{2000\ \text{mL}} \times 100 = 15\%$ Settleable Solids

4. $\dfrac{2.2\ \text{equivalents}}{1.25\ \text{liters}} = 1.8\ \text{N}$

5. $\dfrac{7.5\ \text{mg/L} - 4.2\ \text{mg/L}}{\dfrac{5\ \text{mL}}{300\ \text{mL}}} = (3.3\ \text{mg/L})\dfrac{(300)}{5}$

   $= 198\ \text{mg/L BOD}$

6. (a) <u>Total Solids</u>   <u>Sludge Sample</u>

   27.70 g TS & Dish   80.14 g Sludge & Dish
   − 25.40 g Dish   − 25.40 g Dish
     2.3 g TS   54.74 g Sludge

   $\begin{array}{l}\%\ \text{Total}\\ \text{Solids}\end{array} = \dfrac{2.3\ \text{g}}{54.74\ \text{g}} \times 100$

   $= 4.2\%$ Total Solids

   (b) <u>Volatile Solids</u>

   27.70 g Before Burning
   − 26.14 g After Burning
     1.56 g VS

   $\begin{array}{l}\%\ \text{Volatile}\\ \text{Solids}\end{array} = \dfrac{1.56\ \text{g}}{2.3\ \text{g}} \times 100$

   $= 68\%$ Volatile Solids

7. $\dfrac{80\ \text{grams}}{24\ \text{grams/mole}} = 3.3\ \text{moles}$

8. (a) First determine grams SS:

$$\begin{array}{r} 29.2686 \text{ g}\quad \text{Dish \& SS} \\ -\,29.2640 \text{ g}\quad \text{Dish} \\ \hline 0.0046 \text{ g}\quad \text{SS} \end{array}$$

Then calculate mg/L SS:

$$\frac{0.0046 \text{ g SS}}{50 \text{ m}L} \times \frac{1000 \text{ mg}}{1 \text{ g}} \times \frac{20}{20} = 92 \text{ mg}/L \text{ SS}$$

(b) <u>Volatile SS</u>

$$\begin{array}{r} 29.2686 \text{ g}\quad \text{Before Burning} \\ -\,29.2657 \text{ g}\quad \text{After Burning} \\ \hline 0.0029 \text{ g}\quad \text{VSS} \end{array}$$

$$\% \text{ VSS} = \frac{0.0029}{0.0046} \times 100$$

$$= 63\% \text{ VSS}$$

9. $\dfrac{218 \text{ m}L}{2.31 \text{ g}} = 94 \text{ SVI}$

10. $76° \text{ F} + 40° = 116°$

And $\dfrac{(5)(116°)}{9} = 64°$

Then $64° - 40° = 24° \text{ C}$

11. 7-day average on 10th: (10th + 6 previous days)

$$\frac{207 + 194 + 189 + 196 + 208 + 211 + 205}{7} = \frac{1410}{7} = 201 \text{ mg}/L$$

Average on the 11th: (Using the shortcut method)

$$\frac{1410 - 207 + 212}{7} = \frac{1415}{7} = 202 \text{ mg}/L$$

Average on the 12th: (Using the shortcut method)

$$\frac{1415 - 194 + 202}{7} = \frac{1423}{7} = 203 \text{ mg}/L$$

12. $\dfrac{2190 \text{ mg}}{190 \text{ m}L} \times 100 = \dfrac{2.19 \text{ g}}{190 \text{ m}L} \times 100 = 1.15$

ACHIEVEMENT TEST 18—Cont'd

13. $(0.5)(x \text{ mL}) = (0.1)(800 \text{ mL})$

$$x = \frac{(0.1)(800)}{0.5}$$

$$= 160 \text{ mL}$$

14. (a) <u>Total Solids</u>   <u>Sludge Sample</u>

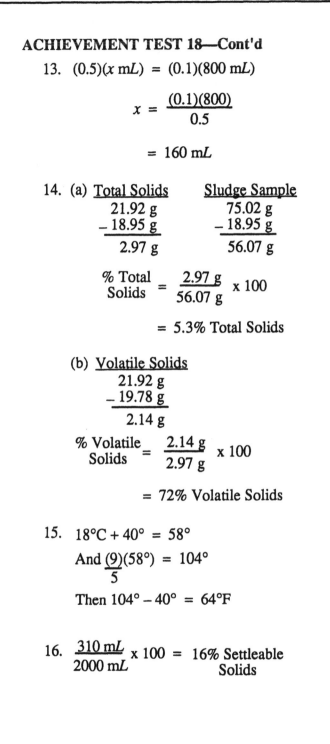

       21.92 g   75.02 g
      – 18.95 g   – 18.95 g
        2.97 g   56.07 g

$$\frac{\% \text{ Total}}{\text{Solids}} = \frac{2.97 \text{ g}}{56.07 \text{ g}} \times 100$$

$$= 5.3\% \text{ Total Solids}$$

(b) <u>Volatile Solids</u>

     21.92 g
  – 19.78 g
     2.14 g

$$\frac{\% \text{ Volatile}}{\text{Solids}} = \frac{2.14 \text{ g}}{2.97 \text{ g}} \times 100$$

$$= 72\% \text{ Volatile Solids}$$

15. $18°C + 40° = 58°$

And $\dfrac{(9)(58°)}{5} = 104°$

Then $104° - 40° = 64°F$

16. $\dfrac{310 \text{ mL}}{2000 \text{ mL}} \times 100 = 16\%$ Settleable Solids

NOTES:

Appendices

FRICTION LOSS IN FEET PER 100-FT LENGTH OF PIPE
(Based on Williams & Hazen Formula Using $C = 100$)

½-in Pipe		¾-in Pipe		1-in Pipe		1¼-in Pipe		1½-in Pipe		2-in Pipe		2½-in Pipe		3-in Pipe		4-in Pipe		5-in Pipe		6-in Pipe		
Vel ft/sec	Loss in ft	Vel	Loss	Vel	Loss	Vel	Loss	Vel	Loss	Vel	Loss	Vel	Loss	Vel	Loss	Vel	Loss	Vel	Loss	Vel	Loss	US gal per min
2.10	7.4	1.20	1.9																			2
4.21	27.0	2.41	7.0	1.49	2.14																	4
6.31	57.0	3.61	14.7	2.23	4.55	1.29	1.20	.94	.56	.61	.20											6
8.42	98.0	4.81	25.0	2.98	7.8	1.72	2.03	1.26	.95	.82	.33	.52	.11									8
10.52	147.0	6.02	38.0	3.72	11.7	2.14	3.05	1.57	1.43	1.02	.50	.65	.17	.45	.07							10
		7.22	53.0	4.46	16.4	2.57	4.3	1.89	2.01	1.23	.79	.78	.23	.54	.10							12
		9.02	80.0	5.60	25.0	3.21	6.5	2.36	3.00	1.53	1.08	.98	.36	.68	.15							15
		10.84	108.2	6.69	35.0	3.86	9.1	2.83	4.24	1.84	1.49	1.18	.50	.82	.21							18
		12.03	136.0	7.44	42.0	4.29	11.1	3.15	5.20	2.04	1.82	1.31	.61	.91	.25	.51	.06					20
				9.30	64.0	5.36	16.6	3.80	7.30	2.55	2.73	1.63	.92	1.13	.38	.64	.09					25
				11.15	89.0	6.43	23.0	4.72	11.0	3.06	3.84	1.96	1.29	1.36	.54	.77	.13	.49	.04			30
				13.02	119.0	7.51	31.2	5.51	14.7	3.57	5.10	2.29	1.72	1.59	.71	.89	.17	.57	.06			35
				14.88	152.0	8.58	40.0	6.30	18.8	4.08	6.6	2.61	2.20	1.82	.91	1.02	.22	.65	.08			40
						9.65	50.0	7.08	23.2	4.60	8.2	2.94	2.80	2.04	1.15	1.15	.28	.73	.09			45
						10.72	60.0	7.87	28.4	5.11	9.9	3.27	3.32	2.27	1.38	1.28	.34	.82	.11	.57	.04	50
						11.78	72.0	8.66	34.0	5.62	11.8	3.59	4.01	2.45	1.58	1.41	.41	.90	.14	.62	.05	55
						12.87	85.0	9.44	39.6	6.13	13.9	3.92	4.65	2.72	1.92	1.53	.47	.98	.16	.68	.06	60
						13.92	99.7	10.23	45.9	6.64	16.1	4.24	5.4	2.89	2.16	1.66	.53	1.06	.19	.74	.076	65
						15.01	113.0	11.02	53.0	7.15	18.4	4.58	6.2	3.18	2.57	1.79	.63	1.14	.21	.79	.08	70
						16.06	129.0	11.80	60.0	7.66	20.9	4.91	7.1	3.33	3.00	1.91	.73	1.22	.24	.85	.10	75
						17.16	145.0	12.59	68.0	8.17	23.7	5.23	7.9	3.63	3.28	2.04	.81	1.31	.27	.91	.11	80
						18.21	163.8	13.38	75.0	8.68	26.5	5.56	8.1	3.78	3.54	2.17	.91	1.39	.31	.96	.12	85
						19.30	180.0	14.71	84.0	9.19	29.4	5.88	9.8	4.08	4.08	2.30	1.00	1.47	.34	1.02	.14	90
								14.95	93.0	9.70	32.6	6.21	10.8	4.22	4.33	2.42	1.12	1.55	.38	1.08	.15	95
								15.74	102.0	10.21	35.8	6.54	12.0	4.54	4.96	2.55	1.22	1.63	.41	1.13	.17	100
								17.31	122.0	11.23	42.9	7.18	14.5	5.00	6.0	2.81	1.46	1.79	.49	1.25	.21	110
								18.89	143.0	12.25	50.0	7.84	16.8	5.45	7.0	3.06	1.7	1.96	.58	1.36	.24	120
8″ Pipe								20.46	166.0	13.28	58.0	8.48	18.7	5.91	8.1	3.31	1.97	2.12	.67	1.47	.27	130
.90	.08							22.04	190.0	14.30	67.0	9.15	22.3	6.35	9.2	3.57	2.28	2.29	.76	1.59	.32	140
.96	.09									15.32	76.0	9.81	25.5	6.82	10.5	3.82	2.62	2.45	.88	1.70	.36	150
1.02	.10									16.34	86.0	10.46	29.0	7.26	11.8	4.08	2.91	2.61	.98	1.82	.40	160
1.08	.11									17.36	96.0	11.11	34.1	7.71	13.3	4.33	3.26	2.77	1.08	1.92	.45	170
1.15	.13									18.38	107.0	11.76	35.7	8.17	14.0	4.60	3.61	2.94	1.22	2.04	.50	180
1.21	.14	10″ Pipe								19.40	118.0	12.42	39.6	8.63	15.5	4.84	4.01	3.10	1.35	2.16	.55	190
1.28	.15									20.42	129.0	13.07	43.1	9.08	17.8	5.11	4.4	3.27	1.48	2.27	.62	200
1.40	.18	.90	.06							22.47	154.0	14.38	52.0	9.99	21.3	5.62	5.2	3.59	1.77	2.50	.73	220
1.53	.22	.98	.07							24.51	182.0	15.69	61.0	10.89	25.1	6.13	6.2	3.92	2.08	2.72	.87	240
1.66	.25	1.06	.08							26.55	211.0	16.99	70.0	11.80	29.1	6.64	7.2	4.25	2.41	2.95	1.00	260
1.79	.28	1.15	.09									18.30	81.0	12.71	33.4	7.15	8.2	4.58	2.77	3.18	1.14	280
1.91	.32	1.22	.11									19.61	92.0	13.62	38.0	7.66	9.3	4.90	3.14	3.40	1.32	300
2.05	.37	1.31	.12									20.92	103.0	14.52	42.8	8.17	10.5	5.23	3.54	3.64	1.47	320
2.18	.41	1.39	.14									22.22	116.0	15.43	47.9	8.68	11.7	5.54	3.97	3.84	1.62	340
2.30	.45	1.47	.15	12″ Pipe								23.53	128.0	16.34	53.0	9.19	13.1	5.87	4.41	4.08	1.83	360
2.43	.50	1.55	.17	1.08	.069							24.84	142.0	17.25	59.0	9.69	14.0	6.19	4.86	4.31	2.00	380
2.60	.54	1.63	.19	1.14	.075	14″ Pipe						26.14	156.0	18.16	65.0	10.21	16.0	6.54	5.4	4.55	2.20	400
2.92	.68	1.84	.23	1.28	.095									20.40	78.0	11.49	19.8	7.35	6.7	5.11	2.74	450
3.19	.82	2.04	.28	1.42	.113	1.04	.06							22.70	98.0	12.77	24.0	8.17	8.1	5.68	2.90	500
3.52	.97	2.24	.33	1.56	.135	1.15	.07							24.96	117.0	14.04	28.7	8.99	9.6	6.25	3.96	550
3.84	1.14	2.45	.39	1.70	.159	1.25	.08							27.23	137.0	15.32	33.7	9.80	11.3	6.81	4.65	600
4.16	1.34	2.65	.45	1.84	.19	1.37	.09									16.59	39.0	10.62	13.2	7.38	5.40	650
4.46	1.54	2.86	.52	1.99	.22	1.46	.10	16″ Pipe								17.87	44.9	11.44	15.1	7.95	6.21	700
4.80	1.74	3.06	.59	2.13	.24	1.58	.11									19.15	51.0	12.26	17.2	8.50	7.12	750
5.10	1.90	3.26	.66	2.27	.27	1.67	.13									20.42	57.0	13.07	19.4	9.08	7.96	800
5.48	2.20	3.47	.75	2.41	.31	1.79	.14	1.36	.08							21.70	64.0	13.89	21.7	9.65	8.95	850
5.75	2.46	3.67	.83	2.56	.34	1.88	.16	1.44	.084	20″ Pipe								14.71	24.0	10.20	10.11	900
6.06	2.72	3.88	.91	2.70	.38	2.00	.18	1.52	.095									15.52	26.7	10.77	11.20	950
6.38	2.97	4.08	1.03	2.84	.41	2.10	.19	1.60	.10	1.02	.04							16.34	29.2	11.34	12.04	1000
7.03	3.52	4.49	1.19	3.13	.49	2.31	.23	1.76	.12	1.12	.04							17.97	34.9	12.48	14.55	1100
7.66	4.17	4.90	1.40	3.41	.58	2.52	.27	1.92	.14	1.23	.05							19.61	40.9	13.61	17.10	1200
8.30	4.85	5.31	1.62	3.69	.67	2.71	.32	2.08	.17	1.33	.06									14.72	18.4	1300
8.95	5.50	5.71	1.87	3.98	.78	2.92	.36	2.24	.19	1.43	.064	24″ Pipe								15.90	22.60	1400
9.58	6.24	6.12	2.13	4.26	.89	3.15	.41	2.39	.21	1.53	.07									17.02	25.60	1500
10.21	7.00	6.53	2.39	4.55	.98	3.34	.47	2.56	.24	1.63	.08									18.10	26.9	1600
11.50	8.78	7.35	2.95	5.11	1.21	3.75	.58	2.87	.30	1.84	.10	1.28	.04									1800
12.78	10.71	8.16	3.59	5.68	1.49	4.17	.71	3.19	.37	2.04	.12	1.42	.05									2000
14.05	12.78	8.98	4.24	6.25	1.81	4.59	.84	3.51	.44	2.25	.15	1.56	.06									2200
15.32	14.2	9.80	5.04	6.81	2.08	5.00	.99	3.83	.52	2.45	.17	1.70	.07	1.09	.02							2400
		10.61	5.81	7.38	2.43	5.47	1.17	4.15	.60	2.66	.20	1.84	.08	1.16	.027							2600
		11.41	6.70	7.95	2.75	5.84	1.32	4.47	.68	2.86	.23	1.98	.09	1.27	.03							2800
		12.24	7.62	8.52	3.15	6.01	1.49	4.79	.78	3.08	.27	2.13	.10	1.37	.037							3000
		13.05	7.8	9.10	3.51	6.68	1.67	5.12	.88	3.27	.30	2.26	.12	1.46	.041							3200
		14.30	10.08	9.95	4.16	7.30	1.97	5.59	1.04	3.59	.35	2.49	.14	1.56	.047							3500
		15.51	13.4	10.80	4.90	7.98	2.36	6.07	1.20	3.88	.41	2.69	.17	1.73	.05							3800
				11.92	5.88	8.76	2.77	6.70	1.44	4.29	.49	2.99	.20	1.91	.07							4200
				12.78	6.90	9.45	3.22	7.18	1.64	4.60	.56	3.20	.22	2.04	.08							4500
				14.20	8.40	10.50	3.92	8.01	2.03	5.13	.68	3.54	.27	2.26	.09							5000
						11.55	4.65	8.78	2.39	5.64	.82	3.90	.33	2.50	.11							5500
						12.60	5.50	9.58	2.79	6.13	.94	4.25	.38	2.73	.13							6000
						13.65	6.45	10.39	3.32	6.64	1.10	4.61	.45	2.96	.15							6500
						14.60	7.08	11.18	3.70	7.15	1.25	4.97	.52	3.18	.17							7000
								12.78	4.74	8.17	1.61	5.68	.66	3.64	.23							8000
								14.37	5.90	9.20	2.01	6.35	.81	4.08	.28							9000
								15.96	7.19	10.20	2.44	7.07	.98	4.54	.33							10000
										12.25	3.41	8.50	1.40	5.46	.48							12000

RESISTANCE OF VALVES AND FITTINGS TO FLOW

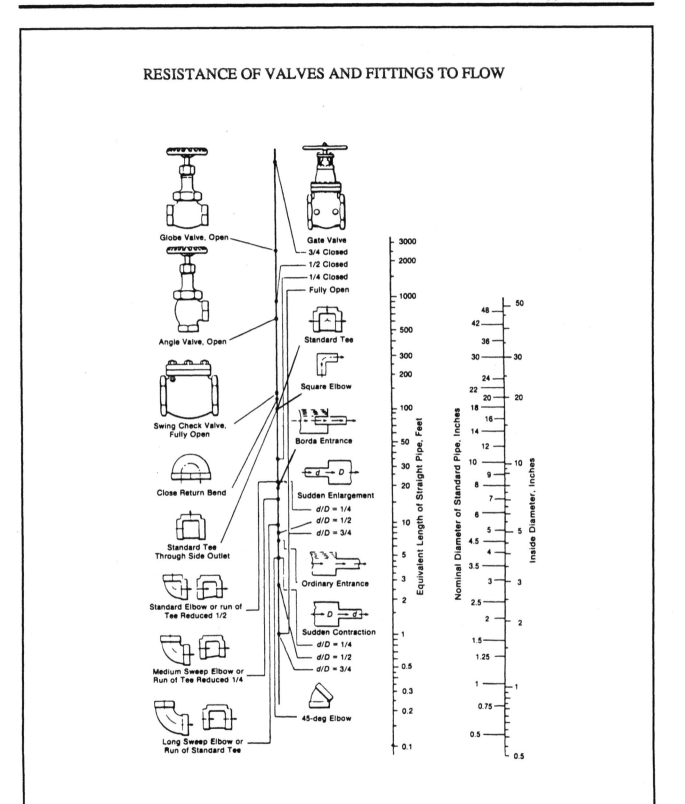

Reprinted with permission of Crane Valves

COMPOST PROCESS CAPACITY NOMOGRAPH*

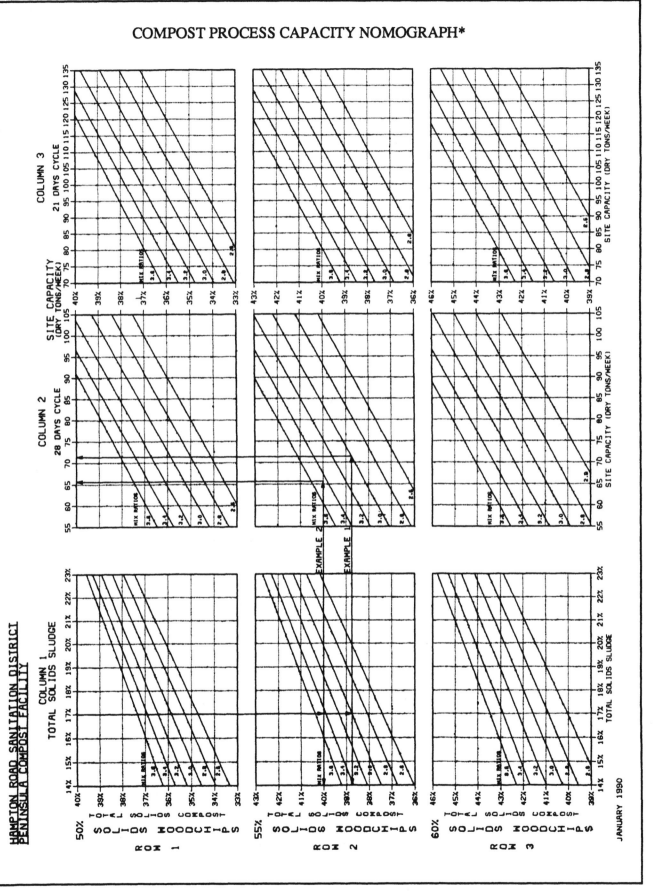

HAMPTON ROAD SANITATION DISTRICT
PENINSULA COMPOST FACILITY

JANUARY 1990

---

* Source: *"Graphical Techniques for Quick and Comprehensive Evaluation of the Compost Process,"* presented at the annual WPCF conference, October, 1990, by Alan B. Cooper, Regional Manager and Senior Project Manager for Black and Veatch, Gaithersburg, Maryland.